Praise for *Am*

"Rasenberger renders 1908 as a series o. . . . . , ... never blinks." —*Publishers Weekly*

"Jim Rasenberger has found the perfect aperture through which to view the explosion of modernity. 1908 was indeed a big bang of a year, a year full of hope and promise but also one which presented our world with a Pandora's box of unforeseen perils. Readers will love—and historians will envy—the graceful simplicity of Rasenberger's singular prism. *America, 1908* effortlessly transports us back to the future, to a distant time and place that seems oddly familiar."

—Hampton Sides, author of *Blood and Thunder* and *Ghost Soldiers*

"This is a wonderful surprise of a book—a time machine back to the year when the American Century got going full tilt. Jim Rasenberger writes in a voice as winning as Theodore Roosevelt's smile and pilots his machine with a sure-handedness that would have impressed the Wright brothers. When you finish *America, 1908,* you will swear you were there."
—Patricia O'Toole, author of *When Trumpets Call: Theodore Roosevelt after the White House*

"An exhilarating panorama of the United States as it was a century ago. The cast of characters here, from Teddy Roosevelt to Fred Merkle (the luckless batter whose mistake lost the New York Giants a still-legendary pennant race), is unforgettable. And the America that shows itself in this masterful narrative constantly reveals links to America today."
—Mark Caldwell, author of *New York Night: The Mystique and Its History*

"*America, 1908* is an intricate time machine with moving parts that mesh like a fine old gold watch, transporting the reader to a time extraordinarily like and yet unlike our own. Rasenberger, a master of detail, gives us a superb rendition of an important and fascinating American moment." —James Tobin, author of *To Conquer the Air: The Wright Brothers and the Great Race for Flight*

ALSO BY JIM RASENBERGER

*High Steel: The Daring Men Who Built
the World's Greatest Skyline*

AMERICA, 1908

The Dawn of Flight, the Race to the Pole, the Invention
of the Model T, and the Making of a Modern Nation

Jim Rasenberger

SCRIBNER
New York London Toronto Sydney

SCRIBNER

A Division of Simon & Schuster, Inc.

1230 Avenue of the Americas

New York, NY 10020

Copyright © 2007 by Jim Rasenberger

First Scribner trade paperback edition June 2011

SCRIBNER and design are registered trademarks of The Gale Group, Inc.,
used under license by Simon & Schuster, Inc., the publisher of this work.

For information about special discounts for bulk purchases,
please contact Simon & Schuster Special Sales at
1-866-506-1949 or business@simonandschuster.com.

The Simon & Schuster Speakers Bureau can bring authors to
your live event. For more information or to book an event, contact
the Simon & Schuster Speakers Bureau at 1-866-248-3049
or visit our website at www.simonspeakers.com.

DESIGNED BY ERICH HOBBING

Text set in Adobe Caslon

Manufactured in the United States of America

3   5   7   9   10   8   6   4

Library of Congress Control Number: 2007025885

ISBN 978-0-7432-8077-8
ISBN 978-0-7432-8078-5 (pbk)
ISBN 978-1-4165-5262-8 (ebook)

For Ann

CONTENTS

PROLOGUE

Leap Year

Any book that hopes to convince its readers that a particular year of America's past was pivotal—that it was "The Year That Changed Everything," or "The Year That Made America," or some such grandiose claim—is suspect at the outset. History does not fit neatly into the annual cycles of the Gregorian calendar. A year is not a historical category, it's a cosmological event: a single revolution of the earth around the sun; three hundred and sixty-five spins on its polar axis; or, in the case of 1908, a leap year, three hundred and sixty-six.

With that in mind, here is a more modest claim: 1908, by whatever quirk of history or cosmology, was one hell of a ride around the sun.

It began at midnight on January 1, 1908, when a seven-hundred-pound "electric ball" fell from the flagpole atop the New York Times building—the first-ever ball-drop in Times Square—and ended December 31, 1908, with a two-and-a-half-hour flight by Wilbur Wright, the longest ever made in an airplane. In the days between, the Great White Fleet sailed around the world, Admiral Robert Peary began his conquest of the North Pole, Dr. Frederick Cook reached the North Pole (or claimed to), six automobiles set out on a twenty-thousand-mile race from New York City to Paris (via the frozen Bering Strait), and the New York Giants battled the Chicago Cubs in one of the strangest and most thrilling seasons in the history of baseball.

Most dramatically, in the spring of 1908, then again in late summer, the Wright brothers proved to the world they'd mastered the skies. The Wrights had first flown at Kitty Hawk in 1903, but not until August and September of 1908, during a series of simultaneous aerial demonstrations at Fort Myer, Virginia (Orville), and Le Mans, France (Wilbur), did America and the rest of the world truly grasp what they had accomplished. That September was, quite simply, one of the most astonishing months in modern history.

A few weeks after Orville Wright's flight trials ended spectacularly (and tragically) at Fort Myer, an automobile called the Model T went into production at Henry Ford's Piquette Avenue plant in Detroit, Michigan. The Model T very soon transformed not only the automotive industry but, in turn, almost every aspect of how Americans lived and worked. Ford's automobile arrived more quietly than the Wrights' airplane, but its impact on the daily lives of most Americans was more immediate and far greater.

Whole books could be written about each of the events that made 1908 remarkable; whole books *have* been written about a few of them. This book considers the events not as discrete moments, though, but as pieces of a larger picture; which is to say, as they were considered by those who lived through them. Whether practically significant, like the launch of the Model T, or essentially frivolous, like the automobile race around the world—an "act of splendid folly," one of the contestants called it—all reflected, and expanded, Americans' sense of what was possible. Only in retrospect would it be easy to distinguish between the credible and incredible. Yes, it was absurd for someone to expect to drive an automobile across the ice of the Bering Strait. But no more absurd, at the start of 1908, than someone proposing to fly an airplane for two and a half hours.

Even barring extravagant claims about 1908, there is no doubt that the America that emerged from the year was different than the one that went into it, and was, in many ways, defined by it for years to come. Buoyed by achievements, the country was more confident in its genius and resourcefulness—not to mention its military might—and more comfortable in assuming its destiny in global affairs. That destiny would be made manifest nine years later, when America entered World War I and changed the course of both the war and the world. For the moment, it was enough to know that America was singularly equipped to lead the world into the twentieth century.

The America in which the following stories occurred will be, in scattered glimpses, familiar to readers of 2008; if we squint, we might even mistake it for our own. There were forty-six states in 1908, nearly the full deck. Oklahoma had just been admitted to the union (1907), and Arizona and

New Mexico were next (1912). Alaska and Hawaii were still half a century from statehood, but the basic geography of the country we know was drawn and settled.

In the cities, middle-class life had assumed some of the shapes and contours of modernity familiar to us. People rode subways to work, lit their homes with electricity, and spoke on the telephone. Many others vacuumed their floors with vacuum cleaners and washed their clothes with washing machines, and all lived surrounded by products whose brand names we recognize: Kellogg's and Nabisco, Coca-Cola and Pepsi, Gillette and Hoover, Budweiser and Pabst, Ford and Buick, among many others.

Though still years away from radio and decades from television, they enjoyed their leisure much as we do. The more affluent disported themselves with tennis and golf. The less affluent preferred baseball and flocked to ballparks in record numbers to watch their champions battle for glory. Movies were very new but already immensely popular; over the previous few years, nickelodeons had sprung up in every small town around the country, from Florida to Alaska.

Americans of 1908 were fascinated, like us, by technology, including wireless technology of the sort being developed at the time by Guglielmo Marconi and Lee De Forest. They anticipated the day when people would walk around with telephones attached to their ears, and when voices and images would whisk through the air across continents and oceans. They paid attention to their diets, read newspaper ads touting advances in plastic surgery, worried that Christmas was becoming too commercialized, that the environment was being ravaged, and that life was speeding up and passing them by. They tried to slow life down by reaching for an Eastman Kodak "Brownie" camera and taking a snapshot.

One other thing Americans of 1908 shared with those of us who live in 2008 was a presidential election. They were coming off two terms of a Republican president who had abruptly set their country on a new course. Like our current outgoing president, theirs was a wealthy Ivy League–educated easterner who had gone west as a young man and made himself into a cowboy. Like ours, theirs had entered the White House without having won the popular vote (in their case, it was an assassination that brought the president into office), then conducted himself with unapologetic swagger. There the comparison between Theodore Roosevelt and George Walker Bush ends; the differences between the two men are far more telling than the similarities. But for the people, it was clear then, as it is now, that the country was heading into a new world defined by as yet unwritten rules, and that the man about to exit office bore some responsibility for this.

Like our respective presidents, we are, for all we hold in common with our predecessors of a century ago, more different than we are the same. For one thing, there were a lot less of them than there are of us. The population of America in 1908 was 87 million, compared to our current 300 million. They looked different than we do, too, and not just in how they dressed and wore their hair. They were, as a group, lighter skinned; nearly 90 percent were Caucasian, and though a recent wave of immigrants from eastern Europe and southern Italy was shading their whiteness to a slightly swarthier hue, they remained blindingly pale. They were younger, too. The average life expectancy for a Caucasian was about forty-nine years; for an African American, thirty-five years. Millions died early of infectious diseases. Another thirty-five thousand died in industrial accidents. Most of these casualties came from the working class and the poor, including the nearly 2 million children who labored in coal mines, steel mills, and other dangerous occupations.

Theirs was a country of extreme divisions and disparities: between the small numbers of the very rich—the 2 percent who possessed 60 percent of the nation's wealth—and the vast ranks of the poor; between whites who were reflexively and almost uniformly racist and blacks who were severely discriminated against (and were forbidden by law in twenty-six states to marry a white person). The division between men and women, too, was stark. The sexes lived in separate realms, if not always in practice then at least in theory. Men worked, women did not; men were strong, women were weak; men enjoyed sex, women did not. That was the theory, anyway, and many Americans believed it.

Finally, no division was more obvious than the one between city and country. Away from the cities, America was still largely rural and pre-modern, lacking electricity or plumbing. More than 50 percent of Americans lived in rural surroundings, and a third of working-age people made their living from the land. They were often informed by magazines and newspapers that they lived in the "Age of Steel" or the "Machine Age," but for most it was still very much an age of horses and cows. There were more than 20 million horses in the country, nearly one for every four people. The single largest industry in the country was not steel or coal or oil. It was meatpacking.

The western states that provided much of the nation's beef were no longer wild frontier by 1908, but they were still largely unpopulated. The demographic median of the country was located near the middle of Indiana, several hundred miles east of the Mississippi River. West of the Mississippi a few mid-sized cities rose from the prairies and the plateaus, but mainly what lay out there was land and those who meant to profit by it. North Dakota and Montana were still open to homesteaders. Texans were indulging in their oil deposits (the famous Spindletop gusher had been discovered in 1901) and Arizona and New Mexico were largely unsettled ranch country. "Now is a good time to come to Demming," urged a New Mexico newspaper called the *Demming Headlight*. "Land surrounding the town can be gotten for almost a song. If you don't get some of it now you will regret it five years hence."

Farther west, Nevada claimed fewer than eighty thousand residents but a gold-mining boom was sucking in new settlers and had nearly doubled the population since 1900. Las Vegas was a dusty depot surrounded by desert ranches. Los Angeles was a city of three hundred thousand people, rich in fruit and oil and overlorded by Harrison Gray Otis, eccentric publisher of the *Los Angeles Times*, who had himself chauffeured around town in an automobile with a cannon mounted on its hood. Otis's fellow Los Angeleno, William Mulholland, was supervising the construction of a 230-mile aqueduct to ensure a steady supply of water for the future city. Mulholland could hardly have imagined what that city would become after 1908, with its green lawns, swimming pools, and four million thirsty citizens, but he went a long way in assuring it.

Which brings us back to the stories in this book. The old frontier had been settled by 1908, but the new frontier was nothing less than the future. Americans of 1908 were confident they could achieve dominion over the future, just as they'd achieved dominion over the frontier of the past. Their confidence made them wonderfully ambitious but a little unhinged. They cherished boundaries as a matter of moral principle— for they were profoundly moralistic people—but they were bursting through every boundary that existed. The last great wave of explorers was setting off to map the last true wildernesses of the world that year,

but the borders that defined human possibility, physical and psychological, political and sexual, were vaguer than ever. What was real and solid
in this strange new world where people could fly and buildings could rise
to the clouds, where submarines plunged into the seas and invisible
words traveled through the air? Where did plausibility end and fantasy
begin?

From certain perspectives, America of 1908 resembled Winsor McCay's
Little Nemo in Slumberland, the immensely popular comic that appeared
inside the *New York Herald* every Sunday. In each strip, a boy goes to sleep
and dreams fantastic adventures, only to wake up in the last panel and
return to reality. Reality? The adventures of Little Nemo were only
slightly less believable than what was reported on the front pages of the
newspapers. According to the press, everything that happened in 1908 was
bigger, better, faster, and stranger than anything that had happened
before. In part, this was newspaper hyperbole; in part, it was simply
true. The world was changing with stunning velocity, as if its very orbit
were accelerating. Life was exhilarating, but also disorienting. Under
the influence of economic, cultural, and demographic shifts, Americans
of all classes and backgrounds found themselves thrashing around in a
state of flux, agitated and edgy, suspicious of each other's motives. The
term "melting pot" entered the American lexicon in the fall of 1908, coined
by playwright Israel Zangwill to define the nation's capacity to absorb and
assimilate different ethnicities and cultures. To our ears, the words may
sound warm and delicious, like a pot of stew, but to Zangwill the melting pot was a cauldron "roaring and bubbling," as he wrote, "stirring and
seething."

In the end, for all that was wrong with America in 1908, the most
impressive trait shared by its people was the hope that carried them
forward. They fiercely believed, not always with good reason, that the
future would be better than the present. It's striking, in fact, how much
more hopeful Americans were then than we are today. We live in a
nation that is safer, healthier, richer, more egalitarian, and less physically
taxing than it was in 1908, but a recent Pew poll found that barely one-
third of us feels optimistic about the future. Upon being asked whether
children growing up in this country would be better off or worse off in the
future than people are today, two-thirds of respondents predicted that they
will be worse off. If this poll accurately reflects our beliefs, the old truism
that Americans are eternally optimistic is apparently no longer true.

Of course, we have reason to be less optimistic. We are wiser now to the downsides of those technologies we've inherited from our predecessors. We cannot look at an airplane without knowing the death and destruction, from Dresden to 9/11, that airplanes have wrought. Automobiles may have once promised their owners exhilarating freedoms, but they now deliver traffic jams, addict us to foreign oil (1908 was the year, coincidentally, that oil was discovered in Iran), and pump doses of carbon dioxide into our atmosphere that will alter the earth in ways few of us dare to imagine. The American pride that sailed with the Great White Fleet on its voyage around the world in 1908, and was met with gratifying adoration at every port, is now tempered by the knowledge that much of the world despises us. The next hundred years may be the price we pay for the conveniences and conquests of the last hundred.

What follows is an attempt to recount the disparate events that Americans of 1908 confronted on their way into their own futures. Confined to a single journey of the earth around the sun exactly one century ago, it is the story of several smaller journeys, beginning, on the first day of the year in New York City, with a boy on a bicycle.

PART I

Dementia Americana

CHAPTER ONE

The Boy and the Machine

NEW YEAR'S DAY

Anything, everything, is possible.
—Thomas Edison, 1908

On the cool, fine afternoon of January 1, 1908, a sixteen-year-old boy named Terrance Kego—or Tego, as several brief accounts had it in the next day's papers—stepped onto his bicycle at his home on West 131st Street and began pedaling down Amsterdam Avenue in the direction of Central Park. Other than his address and his occupation as a clerk, few details about the boy survive. A few more, though, can be surmised.

As he started down the wide avenue, descending from Harlem Heights to the valley at 125th Street, he would have passed through a sloping neighborhood of row houses and low apartment buildings occupied by working-class families. Because today was a holiday, and because the weather was pleasant, some of the families would have been out on the avenue, strolling the bluff above the river. Young children would have turned to watch Terrance glide by on his bicycle, hunched over his handlebars, cap pulled low on his head, wind pulling at the tail of his coat. Perhaps a few flecks of confetti escaped from the furls of his coat and fluttered out behind him like tiny bright moths.

Certainly Terrance had gone out to greet the New Year the previous evening. What sixteen-year-old boy could have resisted the tug of the street? He may have joined the swollen tide of revelers on 125th Street, where the festivities had continued, with occasional interruptions from the police, until nearly dawn. Or more likely, being a self-supporting and spirited adolescent—the kind of go-getter, according to the next day's *New York World*, who had made a New Year's vow to "take no one's dust when on his bicycle"—he'd traveled downtown to Forty-second Street to cast himself into that great cauldron of humanity that was Times Square on New Year's Eve.

Only a few years earlier, well within Terrance's young memory, New Year's Eve had been a quiet and civilized affair spent at home or on the streets of lower Broadway, where the chimes of Trinity Church rang harmoniously at midnight. These last several years, though, it had metamorphosed into something entirely different—more like an election night bacchanalia, with a bit of Independence Day bumptiousness thrown in, plus some frantic energy all its own. The chimes still rang at the old church downtown, but the action was uptown now, and its pulsating center was right here at the nonsectarian intersection of Broadway and Forty-second street.

Arriving in Times Square, Terrance would have climbed directly into a press of bodies and a blizzard of confetti swirling under the dazzling lights of Broadway. The streets had been filling since early in the evening, tens of thousands of bodies funneling in from Union Square and the Flatiron district, from the Tenderloin, still others from the outer boroughs by streetcar or subway or ferry. "An acrobat could hardly have managed to fall down for a wager, so tightly did the people hold each other up," reported the next day's *New York Evening Sun*. A special correspondent for the *Chicago Daily Tribune* judged the noise in Times Square to be more varied than in previous years. "Slide trombones that yowled like a cat in torture, a combination of cowbells and street car gongs, tin horns with a double register, sections of iron pipe that could be rasped with files till they gave forth bellows that carried for blocks," were a few of the sounds the correspondent recorded. Shouts and squeals blended with these other sounds to create, as the *New York Tribune* put it, a "terrifying reverberation."

To step into that crowd was to release all sense of direction and decorum. It moved as an organic, unruly mass, drifting, lulling, then surging spasmodically. A sixteen-year-old boy on the last night of 1907 would have been astonished to find himself squeezed in among so many strangers; or, more to the point, among so many young women. While the usual distaff armor—overcoats, ankle-length skirts, petticoats, shirtwaists, steel-plated corsets, undergarments—did its job of keeping feminine flesh secured, the rules of Victorian modesty lapsed that night. Men and women ground against each other indiscriminately.

At every corner, meanwhile, temptation beckoned in the form of vendor carts stocked high with horns and nickel bags of confetti. Assuming you could get to one of these vendors, you could stick a horn into your mouth and stuff your pockets with confetti, all for a dime, then dash back into the swarm to discharge your colored specks into the face of a stranger. More aggressive boys and men squirmed through the crowd with small feather dusters—"ticklers," they were called—brushing the exposed flesh of women's faces and ears, then vanishing before they could be reprimanded or cuffed. Police Commissioner Bingham had issued an edict against the use of ticklers this New Year's Eve but nobody paid him any mind. Many of the women had taken matters into their own hands, covering their faces with heavy veils to ward off the feathers.

Near Forty-second Street, a group of enterprising young men pulled a clothesline across the sidewalk, tying one end to an iron post and drawing the other end taut to the curb. When young women approached, they raised the line a foot or two off the pavement. "Leap year, ladies!" they called. "Take the jump. Show what you can do for leap year!" Some women stepped down onto the street to walk around, but many accepted the challenge. They lifted their skirts above their ankles and jumped.

Young men executed most of the pranks that night, but the females were hardly blameless. Groups of them clasped hands to each other's shoulders to form daisy chains, then ran into the crowd, whipping through it with merry violence. Late in the evening, at Forty-sixth Street and Broadway—was Terrance there to see it?—a dozen young women encircled a well-dressed young man. Locking their arms together, they refused to let him escape. The young man repeatedly tried to bash his way out of their circle but the young women only pushed him back into the center, taunting and jeering him. A policeman finally rescued the hapless young man, but not before the women had kicked his silk hat down Broadway.

What Terrance could not have witnessed that night were the diamond-clad women smoking cigarettes inside the glittering precincts of Rector's and Martin's. These two Broadway restaurants, among others, had relaxed their usual restrictions and allowed female patrons to indulge in tobacco, a fact so remarkable it was recorded on front pages of newspapers around the country. Much as Americans would tune their television sets to

watch the ball drop in Times Square in later years, they looked to New
York that New Year's Eve for excitement, dismay, and provocation. On the
night of the first-ever ball-drop, New York did not disappoint.

Just before midnight, a hush fell over the crowd. All eyes rose to the top
of the Times building. Up there, hovering over the city beyond the glare
of powerful searchlights, poised at the tip of the seventy-foot flagpole, a
giant glittering sphere waited to fall. It was five feet in diameter, seven
hundred pounds in weight, and cloaked in 216 white lights. Nobody
could have guessed they were about to witness the debut of a custom that
would still mark the New Year a century later; nor was it likely, at this
moment, that anyone was peering so far ahead. It was enough, in the
remaining seconds of 1907, to contemplate the difficulties of the year
behind and the promises of the year ahead.

The crowd began to count backward: tens of thousands of voices rising to the sky above New York, joined together in anticipation of something new and marvelous. Then the gleaming orb fell and bright white
numbers flashed on the roof of the Times building: *1908: 1908: 1908.*

At the end of the long glide down to 125th Street, Terrance could either
pedal furiously to gather speed, then take his chances dodging whatever
traffic might be passing along 125th Street—thereby preserving his
momentum for the long ascent to Morningside Heights—or he could
veer east on 125th Street. The latter was the easier, more sensible route.
The thoroughfare would have been quiet on New Year's Day. A few horses
would be standing at the curb before their carts, snorting patiently. An
automobile might rumble past, but automobiles were still scarce in
Harlem, since not many people living on fifteen or twenty dollars a
week could afford one. Other than the piles of horse manure, which Terrance would be mindful to dodge, the street promised a smooth ride over
macadam.

A few effortless blocks, then Terrance would turn south again, skirting Morningside Park. To his right, across the park, rose the stony cliffs
of Morningside Heights. Atop them, blotting the weak afternoon sun,
loomed the gray walls of St. John the Divine, the great cathedral begun
the year Terrance was born and still in the infancy of its construction.
The plain of Jewish Harlem spread out to the east. The road was flat
all the way south to Central Park. Terrance's legs would still be fresh
when he got there.

Had there ever been a finer time to be an American boy on a bicycle than on that first day of the new year of 1908? Certainly the interests and passions of a sixteen-year-old had never coincided so perfectly with those of his country. America was very much an adolescent itself, brash and exuberant, stirring with strange and urgent new longings, one moment supremely confident and clever, the next undone by giddiness and hormones. The psychologist G. Stanley Hall had recently coined the term "adolescence" to describe the passionate "new birth" that occurred in humans between childhood and maturity. It was, wrote Hall, a phase characterized by "storm and stress," but also by joy and delight, as "old moorings were broken and a higher level attained." The description fit the America of 1908 as well as it did any teenager.

Mainly, what a sixteen-year-old American boy and his country shared was a sense of their own glorious futures. Only a few months earlier, in October of 1907, a financial panic—"the flurry," the papers insisted on calling it—had sent stock prices tumbling and thrown millions of Americans out of work. Already, though, the economy seemed poised to rebound and resume its relentless growth. To an ambitious boy already in possession of a bicycle and a job, the future was limitless.

This morning, the *World* had published an essay to greet the New Year in which the paper's editors looked back to the past, then ahead into the future. The title of the piece was simply "*1808—1908—2008.*" The *World* began by noting how far the country had progressed over the previous century. In 1808, five years after the Louisiana Purchase and two years after Lewis and Clark returned from their transcontinental journey, the country's population was a mere seven million souls. The federal government was underfunded and ineffectual. The state of technology—of transportation, communication, medicine, agriculture, manufacturing—was barely more advanced than during the Middle Ages of Europe.

Now, in 1908, with the population of America at almost ninety million, the federal revenue was forty times greater than it had been a century earlier and America was on a par with Britain and Germany as a global power. U.S. citizens enjoyed the highest per-capita income in the world and were blessed with the marvels of railroads and automobiles, telegraph and telephone, electricity and gas. Banks of high-speed elevators zipped through vertical shafts of the tallest buildings on earth. Pneumatic tubes whisked mail between far-flung post offices in minutes.

Men shaved their whiskers with disposable razor blades and women cleaned their homes with remarkable new devices called vacuums. Couples danced to the Victrola in the comfort of their living rooms and snuggled in dark theaters to watch the flickering images of the Vitagraph. Invisible words volleyed across the oceans between the giant antennas of Marconi's wireless telegraph, while American engineers cut a fifty-mile canal through the Isthmus of Panama.

From the glories of the present, the *World* turned to the question of the future: "What will the year 2008 bring us? What marvels of development await the youth of tomorrow?" The U.S. population of 2008 would be 472 million, predicted the *World.* "We may have gyroscopic trains as broad as houses swinging at 200 miles an hour up steep grades and around dizzying curves. We may have aeroplanes winging the once inconquerable air. The tides that ebb and flow to waste may take the place of our spent coal and flash their strength by wire to every point of need. Who can say?"

Not a day passed without new discoveries and advances achieved or promised. This very New Year's Day Dr. Simon Flexner of the Rockefeller Institute was publicly declaring his conviction (in a medical paper read at the University of Chicago) that human organ transplants would soon be common. Meanwhile, the very air seemed charged—it was, actually—with the possibilities of the infant wireless technology. "When the expectations of wireless experts are realized everyone will have his

One of the "possibilities of wireless" as imagined by *Harper's Weekly*, February 1, 1908.

THE PORTABLE WIRELESS TELEPHONE
"TALK WHILE YOU WALK"

own pocket telephone and may be called wherever he happens to be," *Hampton's Magazine* daringly predicted in 1908. "The citizen of the wireless age will walk abroad with a receiving apparatus compactly arranged in his hat and tuned to that one of myriad vibrations by which he has chosen to be called. . . . When that invention is perfected, we shall have a new series of daily miracles."

As a New Yorker, Terrance would have been keenly aware of how abruptly daily miracles were transforming the modern world. The city provided an education in rapacious progress, evolving at the jumpy speed of a Vitagraph picture, escalating toward some yet unwritten but marvelous destiny while accompanied by the tune of pounding rivet guns and 310,000 jangling telephones—six times more phones than in Paris, the New York Telephone Company bragged, and twice as many as in London. Though London remained the largest city in the world, New York was a close second, and twice as large as its nearest American competitor, Chicago. And still growing by leaps and bounds.

"The great city of New York is in that stage of its development in which it may be likened to a youth who is neither man nor boy," observed the *New York Tribune* later that January. "It is in the transition stage. It has not yet lost the ungainliness and awkwardness of the growing youth, but every day can see fresh evidence of its coming symmetry, strength and beauty."

The population of the city had risen by one million inhabitants since Terrance turned nine and continued to grow at a rate of 130,000 per year. In 1907, the largest influx of immigrants in the nation's history—1.29 million—had arrived on American shores, most of them sluicing through New York's Ellis Island, and many staying to settle in East Side tenements and add themselves to the enormity of the city. As the population grew, the city's infrastructure expanded at a mind-boggling rate. Terrance was twelve when the longest and most intricate subway system in the world opened under the streets of New York. A year before that, the Williamsburg Bridge had opened, and since then two new remarkable bridges, the Manhattan and the Queensboro, had been rising over the East River. Tunnel workers were meanwhile blasting and digging beneath both the East and North (Hudson) Rivers. Within the next two months of 1908, two tunnels would open under the rivers, one to Brooklyn, the other to New Jersey. Back aboveground, the Singer Building, the tallest skyscraper in the

THE COSMOPOLIS OF THE FUTURE. A weird thought of the frenzied heart of the world in later times, incessantly crowding the possibilities of aerial and inter-terrestrial construction, when the wonders of 1908—the Singer Bldg., 612 ft. high, with offices on 41st floor, and the Metropolitan Life tower, 658 ft. high—will be far outdone, and the 2,000-ft. structure realized; now nearly a million people do business here each day; by 1930 it is estimated the number will be doubled, necessitating tiers of sidewalks, with elevated lines and new creations to supplement subway and surface cars, with bridges between the structural heights. Airships, too, may connect us with all the world. What will posterity develop?

In his *King's Views* of 1908, publisher Moses King dreams
of the "Cosmopolis of the Future."

world, was nearly complete downtown. Already ironworkers were rais-
ing the steel of a *taller* building, the seven-hundred-foot Metropolitan
Life tower on Madison Square.

Like giant fingers, these towering new structures pointed insistently
to the sky. Railroads and automobiles had penetrated the terrestrial interior, subways and tunnels and mine shafts had delved into the subterranean, and now man would conquer air. Architects and visionaries
imagined future metropolises of one-thousand-foot towers connected by
webs of high-altitude bridges and catwalks; of civilizations of airborne
people who would live above the earth and traverse the crowded skies in
various aircraft. A headline in that morning's *Times* echoed the predic-
tion: MANY MAY FLY IN 1908. The newspaper quoted the French aviator
Henri Farman: "Twelve months from now we will have aeroplanes
which will fly ten or twelve miles easily without once touching the
earth." It was an absurd, fantastic proposition, and yet—

Anything, everything, was possible. If you doubted Thomas Edison's con-
fident assessment, you might look to Harry Houdini for confirmation.
The famed illusionist lived with his wife and mother at 278 West 113th
Street, just down the block from where Terrance was passing now. Inside
his handsome brownstone Houdini was putting the finishing touches on
his latest act, the Milk Can Escape, for a late January debut in St. Louis.
But even Houdini's Milk Can Escape paled next to the marvel of flying.
Nothing, really, compared to flying. (A point Houdini himself would con-
cede when he purchased an airplane a few years later.) Most Americans
still considered airplane flight more of a fantastical dream than a practical reality in these early hours of 1908. The names of Wilbur and Orville
Wright were only vaguely familiar, and those who knew of them had little reason to assume these taciturn bicycle mechanics from Dayton,
Ohio, would emerge as victors in the race to the sky.

How strange, then, that before the end of the year the Wrights would
be more celebrated than Houdini—more celebrated, in fact, than just
about anyone on earth. And the boy, who turned and leaned now,
swooping across 110th Street to the park entrance, would learn the story,
along with every other American, of how Wilbur Wright discovered the
secret of aerial equilibrium by riding a bicycle.

Terrance would have entered the park from the northwest corner, turning east under the bare trees and joining the automobile traffic on the

six-mile loop of the Central Park Drive. Not long ago, the park had been restricted to bicycles and carriages, or, on snowy winter days, to horse-drawn sleighs. Within the last several years, automobiles had come to dominate the drive, and now they roared by Terrance, their engines whining and smoking. Inside the open-bodied cars, parties of "automo-bilists" enjoyed a bracing holiday lark. The women sat snug in squirrel and ermine, their pink cheeks obscured by veils; the men hunkered in masculine bearskins and wore thick goggles over their eyes.

Automobiles had been available in the country for ten years but were still largely vehicles of sport rather than of utility. And with newly acquired speeds of nearly a mile a minute, they assured an experience far more invigorating than any provided by a horse or bicycle or sleigh. By 1908, speed limits were in effect on city streets but drivers tended to ignore them. Speeding, or "scorching," was an integral component of the sport.

An ad for Dewar's scotch in *Harper's Weekly* succinctly captured the devil-may-care attitude of drivers in the early part of the century. An automobile soars over a hill as a gleeful ménage frolics within the open chassis. One of the passengers is reaching back over his shoulder into a basket, pulling out a bottle. "There is no more exhilarating sport or recreation than automobiling," reads the ad copy. "The pleasure of a spin over country roads or through city parks is greatly enhanced if the basket is well stocked with Dewar's Scotch 'White Label'."

Over two hundred thousand cars would ply American roads by the end of 1908, most produced by the five hundred or so American automobile manufacturers that had sprung up since 1901. The vast majority of these companies would soon go out of business, devoured or made obsolete by larger companies. For the moment, though, automobiles came in an array of Whites and Waynes and Haynes, of Stearns and Moras and Bakers, of Pierces and Popes and Thomases—an endless list of eponymous entrepreneurs who had seen the motorized future but would have no place in it. They came in every shade and shape, these automobiles, but the fashionable color was red and the prevailing size was large. As *Collier's* humorist George Fitch would describe it later in the year, the ideal automobile of 1908 included "room for seven people, a whist table, half a ton of lunch, a cord of golf sticks, a barrel of gasoline, half a peck of small bills for fine money, and a kit of embalming tools in case of emergency."

The embalming tools, *Collier's* readers understood, were to handle the mess of human casualties that automobiles routinely strewed in their wakes. If automobiling was still sport, it was clear by the start of 1908 that it was an exceedingly dangerous sport. "We have become so perfectly accustomed to being whirled along at the rate of sixty miles an hour in our luxuriously appointed motors and private cars that we fail to recognize the positively death-dealing agents of destruction which we are driving through thickly populated streets or along narrow country roads," warned a *New York Times* editorial in January of 1908. Later in the year, *Scientific American* provided an alarmingly opaque explanation of the forces that made automobiles so lethal:

> The kinetic energy which a projectile possesses in virtue of its mass and velocity is necessarily expended in destructive action when the flight of the projectile is suddenly arrested by an obstacle of any sort. The destructive effects are divided between the projectile and the obstacle in the inverse ratio of their respective powers to resist deformation.

All of which seemed to mean that when an automobile came into contact with a smaller object, the smaller object was doomed.

Not everybody considered automobiles health hazards. Indeed, many physicians and health officials looked to them as a salvation from the grime and disease that plagued cities. Not only did they offer the benefits of a restorative drive in the fresh air of the countryside, but they also promised cleaner air within the cities. To understand how anyone could have believed automobiles a *remedy* to air pollution, it helps to recall that more than 120,000 horses tramped the streets of New York in 1908, and that each of these horses dropped about 22 pounds of manure a day, for a total daily deposit of over 2,640,000 pounds. When the manure dried, it drifted through the air as a thick dust that infected nasal passages and respiratory tracks. A writer in *Appleton's* magazine blamed twenty thousand deaths a year on horse dung in 1908. The automobile, he wrote, would overcome "the absurdities of a horse-infected city."

Whatever their impact on air pollution and public health, the real disagreement about automobiles was driven more by economic status than by environmental concerns. Automobile owners tended to be wealthy as a matter of tautology. Since most models cost between two and four thousand dollars (two to four years of salary for one of the working men who lived in Terrance's neighborhood), only the well-off could afford to buy them. The fact that the machines brought out the worst excesses of the

rich, confirming what many Americans already believed about them—
they were callous, feckless, selfish, and ridiculous—added to the resentment of those who could not afford them. "Nothing has spread socialistic
feeling in this country more than the use of the automobile, a picture of
the arrogance of wealth," declared Woodrow Wilson, the president of
Princeton University, in 1906, airing a view that was as reasonable as it was
wrong.

Wilson, like many others, failed to take into account the ripening
attraction of the automobile, fatal or not, to *all* classes of Americans.
Before the end of 1908, a forty-five-year-old Michigan native named
Henry Ford would begin manufacturing a revolutionary new machine
that would be so affordable, so dependable and useful, that few Americans of any class would be able to resist its charms. By the time Wilson
became president of the United States in 1912, even socialists would be
driving Model Ts.

Terrance was pedaling hard now through the northern reaches of Central Park, the grade of the drive rising under him in a long, ascending S.
The automobiles would be chugging past him, their tires spewing bits of
oiled gravel, their engines venting smoke. In the gloaming of late afternoon, a few drivers had probably turned on their glaring acetylene head-
lights to better see the way under the trees. Terrance, the boy who vowed
to take no one's dust, would rise from the seat of his bicycle, tilting forward, his legs pumping and his skin prickling.

Of course, it was impossible for him to overtake an automobile on the
rise of the steepest hill in Central Park; then again, *who's to say?* Maybe
he convinced himself he could do it; maybe he focused intently on the
tonneau of the automobile in front of him and decided to show its occupants a thing or two about speed. Inside the car, a chauffeur was at the
wheel. A family sat in the back. From the corner of his eye, Terrance
might have gauged that they were wealthy even by the standards of
automobile owners. He would have noticed the haughty prep school
boys, home from Groton for the holidays, and the daughters sitting
primly in ermine. One of the girls was fifteen. Maybe she turned and her
eyes met his, issuing a challenge he couldn't refuse.

What happened next is unclear. News accounts vary. In one, the automobile confronted heavy traffic and abruptly stopped. In another account,
the chauffeur, changing lanes to pass another car, pulled out in front of

Terrance. "As it was slowing down the Kego boy crashed into the tonneau," is how the *Times* reported it. The *Herald* was slightly more graphic: "Running into the rear of the automobile, Tego was thrown forward, striking the car." The *World,* typically the most colorful of the lot, painted a fuller, though not necessarily more accurate, picture: "Tego put on steam, and as he rounded a turn, crashed into the rear wheel of the auto and was hurled to the ground." The result, in any case, was the same: Terrance Kego, or Tego, lay unconscious at the rear of the automobile.

The back door of the automobile opened and a man with a moustache and bowler stepped out. The man was no longer young but not so old as to require the cane he customarily carried. He was largely built, but soft, almost diffident, as he set his foot onto the macadam. He might have looked vaguely familiar to other automobilists who witnessed the accident and paused to rubberneck. His name was J. P. Morgan Jr., or simply Jack to those who knew him. Jack Morgan was a man who turned few heads in his own right. His father, though, was among the most celebrated men in the world—and never more celebrated than now. In the wake of the financial panic a few months earlier, J. P. Morgan Sr. had been credited with single-handedly saving Wall Street and restoring faith in the American economy. Even before this recent triumph, numerous regal titles had been attached to him. He was the Duke, the Zeus, the Pope of Wall Street. These days, most often, he was simply the King.

Like his larger-than-life father, Jack Morgan was a tall, burly man. His facial features were similar to his father's, too, though he lacked his father's fantastically florid nose and piercing dark eyes. He lacked his father's iron will and imperious temperament, too. He had lived in the great man's shadow for so long that he was, at forty-one, something of a shadow man. He was also, though, a millionaire many times over. And if his father was King, then he, only son and heir apparent of the House of Morgan, was de facto prince.

Jack Morgan walked over to the boy, stooping slightly, as he always did. Terrance lay unconscious on the ground, his eyes closed but his chest heaving in shallow breaths. With the help of his chauffeur, Jack Morgan carried the boy to the roadside and laid him down. Then, according to the New York *Press,* he did something his regal father would never have done: "The millionaire knelt down and placed the boy's head on his knee." Together, they waited for help to arrive.

Internal Combustion

All the remarkable new possibilities available to Americans at the start of 1908 did not undo some hard realities. When a boy on a bike collided with an automobile, the result was—as *Scientific American* had explained—predictably bad for the boy on the bike. And if, metaphorically speaking, America at the start of 1908 was a boy with a full heart and high hopes, the country was also a kind of pitiless machine.

Machine metaphors were fashionable in the early twentieth century. A few months before Terrance took his New Year's Day bicycle ride, the writer and historian Henry Adams had self-published a book, *The Education of Henry Adams,* in which he described a revelation he'd experienced standing before a row of dynamos at the Paris Exposition of 1900. These gargantuan electricity-generating machines overwhelmed Adams. "To him, the dynamo itself was but an ingenious channel for conveying somewhere the heat latent in a few tons of poor coal hidden in a dirty engine-house carefully kept out of sight," wrote Adams, describing his thoughts in the third person. "As he grew accustomed to the great gallery of machines, he began to feel the forty-foot dynamos as a moral force, much as the early Christians felt the Cross." Adams had stumbled upon what was for him a perfect representation of the churning, gnashing power of the modern age that made people like him feel painfully obsolete. "At seventy, it is hard not to take one's helplessness seriously," he wrote to his old friend Supreme Court Justice Oliver Wendell Holmes on New Year's Eve, 1907—the day before Terrance's ride in the park.

It is fitting that the device Adams used to convey his obsolescence was already a bit fusty by the time he published the *Education.* Electricity had been a fact of life in America since 1882, the year Thomas Edison wired lower Manhattan with direct current. Urban Americans were taking it more or less for granted by the early 1900s. Dynamos were impressively large and mesmerizing in their power, but compared to other rapidly advancing technologies of the early twentieth century, they were dinosaurs. The interest of Americans had been captured by a newer, smaller technology of the internal combustion engine.

The internal combustion engine was (and still is) the hidden mechanism that allowed automobiles to move. About 10 percent of automobiles ran on steam or electricity in 1908, but the gasoline-powered engine was the clear future of automotive technology. It was also the engine that had sent the Wright brothers aloft in 1903 and would, in short order, allow humans to fly through the air at will. The internal combustion engine had not been invented in America, but by 1907 had already proved to be the most significant technological innovation in American history.

Its metaphorical possibilities were as potent as its physical properties. An internal combustion engine, like any engine, is a transformative medium, ingesting one kind of energy—in this case, the dormant explosiveness of gasoline—and releasing another. As gasoline detonates in the confines of a compression chamber, the force of its explosion pushes a piston, which turns a crank shaft, which spins the wheels. Energy is not created; it is converted from a fuel into a force.

America at the start of the twentieth century was such an engine of conversion. Certain forms of energy went in; other forms came out. The fuel that entered the machine was of course the natural resources cut and dug from its land—crops, timber, coal, petroleum—but it was also the human labor that harvested these resources and fed them into the machine: the millions of American men and boys toiling in coal mines and steel mills, in cotton and tobacco fields; the women and girls in sweatshops and textile factories. The working poor gave nearly all their waking lives to labor, ten or twelve hours a day with only Sundays off, for $12 or $15 a week.

In addition to being physically demanding and low paying, much industrial work was extremely dangerous. Altogether, about thirty-five thousand American workers died on the job annually in the first decade of the twentieth century—about ninety-six per day—and another half a million were injured, giving America one of the highest worker-accident rates in the world. Adding an all too literal twist to the machine metaphor, laborers were sometimes consumed, fuellike, by acts of combustion. In steel mills, molten pig iron slopped over the lips of casks, or casks fractured, flooding mills with flaming liquid metal and roasting men alive. In coal mines, gases trapped in deep shafts exploded, igniting, in turn, the coal dust floating through the air. Over three thousand miners had been killed in American mines in 1907. December alone had been astonishingly deadly. Four accidents—two of these in Pennsylvania, another in Alabama, another in West Virginia—had killed 694 miners. In the last accident, on December 19 in Monongah, West Virginia, 362 men and boys had perished; to this day, Monongah remains the worst mining disaster in U.S. history.

Naturally, workers sometimes balked at performing difficult, dangerous work for long hours and low wages. Occasionally they organized strikes, but strikes tended to go poorly for workers, requiring them to endure weeks or months without pay only to find themselves unemployed in the end. A man unable or reluctant to do his job was easily replaced by another man happy to fill his shoes for less money and with less complaint. Generally, this other man was one of the immigrants pouring into America at a rate of thirty-five hundred a day.

Immigrants did not get much respect or affection from native-born Americans. On the contrary, they were roundly disparaged as filthy, criminally inclined, and genetically inferior. But it was they, the foreign-born workers who made up 60 percent of the nation's heavy-industry labor force, who were the secret weapon of American industry. They kept wages low, worked diligently and willingly, and, unlike the nation's coal or iron reserves, were apparently inexhaustible.

Into the engine, then, went inexpensive labor, and out came: wealth. Tremendous, unfathomable wealth. Wealth on a scale enjoyed by few

"Young Driver in Mine" by Lewis Wickes Hine, 1908. Hine quit a teaching job in 1908 to devote himself to photographing child laborers such as this boy.

humans in history. A portion of this wealth was distributed throughout the population as wages and consumer spending, giving Americans the highest per capita income of any nation on earth. Much of it, though, was funneled directly into the hands of the tiny minority who had a stake in the profits of the country's chief industries; the 2 percent of the population, that is, who owned 60 percent of America's wealth.

Then, as now, the rich found ever grander, ever stranger ways to display their wealth. The social critic Thorstein Veblen had earlier described the spending habits of the rich as "conspicuous consumption." A younger critic of the rich, Upton Sinclair, fresh off the great success of his 1906 muckraking novel *The Jungle*, wrote vividly of such consumption in his 1908 follow-up, *The Metropolis*. Sinclair's characters squander astonishing sums of money on fashionable clothing, drive automobiles at reckless speeds, stuff themselves with sickening feasts, divert themselves with "freak fetes"—lavish parties that often featured bizarre costumes and entertainments—and utter such observations as, "The trouble with poor people, it seems to me, is that there are so many of them."

Sinclair overlooked one curious custom of the rich that year. This was the performance of so-called "tableaux." To stage a tableau, a wealthy socialite would invite several other wealthy socialites to dress up as famous historical figures from paintings. The socialites would appear on a stage before an audience of peers, holding significant and attractive poses in the footlights. In one tableau that winter of 1908, arranged by Mrs. Waldorf Astor at the Plaza Hotel, Mrs. James B. Eustle presented herself as Flaubert's Salammbô. She wore a blue sapphire robe and a forty-five-pound live boa constrictor draped over her shoulders. With admirable discipline, she maintained her pose without flinching. The only thing moving was the snake.

It is tempting to read secret yearnings into these taxidermic theatricals: a desire for stillness dressed as history; a wish to freeze time amid the hurly-burly of the modern city. Life for the rich was not all a drive in the park. They struggled with the knowledge that their financial status was tenuous and contingent on the relentlessly dynamic American economy. The 1907 panic served to remind them that wealth could be snatched away in a few bad weeks on Wall Street.

So as they sat in the Plaza ballroom in silence and watched the snake twist around the torso of Mrs. James B. Eustle, they had to wonder how, having achieved position at the very top of the American pyramid, they could be sure to keep it. How might they guarantee that their perch in the social order remained as it was? They required a certificate of status more permanent than anything America could confer, the kind of status only European nobility enjoyed. They required, in other words, *titles*. Hence, the peculiar institution known as the "international marriage." By the terms of international marriage, a rich American heiress would be married off to a less rich European noble. He got wealth, she got title, and everybody was happy. Or, more likely, not so happy, as the unions continued to be doomed. "Sociologists and statisticians have been wasting data to demonstrate that marriages of this kind are invariably productive of misery," mused the *Saturday Evening Post* in its first issue of 1908.

Despite such warnings, the practice showed no signs of abating. The young heiress Gladys Vanderbilt (whose cousin Consuelo was already in a failing international marriage to the ninth duke of Marlborough) was to be married to a Hungarian count at the end of January of 1908. The daughter of Senator Elkins of West Virginia was secretly engaged—or so it was not-so-secretly reported in the press—to the duke of Abruzzi. In February, another heiress, Theodora Shonts, was to marry the penniless duc de Chaulences. Two months after that wedding, the duc would die in a Paris hotel room from an overdose of cocaine and morphine.

And on it went, misery or no. Every day, thousands of poor European immigrants arrived in New York Harbor in steerage to begin their new lives of laboring at a dollar or two a day. And every few weeks, aboard one of those very same Cunard or White Star liners—only now, of course, several decks above in first class—out went the product of their labor, back to the European continent in the pampered and bejeweled form of a newly minted duchess or countess. Upon these great ocean steam liners, coming and going, the American machine achieved its most pure expression.

In the Park

Darkness fell on the first day of 1908 in New York City. Twenty minutes had passed since the boy on the bicycle rammed into the back of the automobile in Central Park, but still no ambulance came for the boy. Terrance remained unconscious, his head resting on Jack Morgan's knee. The accident that brought them together was already dimming in the shadows. The next morning, four newspapers would carry brief items about the accident, each reporting it slightly differently.

Given that accounts disagree on the name of the boy, as well as on the precise location and circumstances of the accident, it's fair to wonder at the accuracy of the whole story. The facts do not quite add up. If the automobile and the bicycle were both climbing the longest, steepest hill in Central Park, how did a sixteen-year-old boy riding a simple-gear bicycle gain enough speed, and therefore sufficient force, to slam into the back of an automobile hard enough to knock himself out? No one who has gasped up this hill on an eighteen-speed twenty-first-century bike will easily understand.

At the risk of impugning the American press corps circa 1908, it should be pointed out that newspaper owners sometimes adjusted their news to suit J. P. Morgan Sr., on the grounds that the financier might pull their operations out from under them if they displeased him. There is a reasonable chance, in other words, that it was not the boy who hit the car but the car that hit the boy, and that the facts of the case were made to fit the laws of finance more accurately than they fit the laws of physics.

Then again, maybe it happened just as it was told. Maybe the boy really did climb the hill so fast, with such speed and determination, that he really did slam into the back of the car with enough force to throw himself unconscious to the ground. *Anything, everything, is possible.*

A police deputy finally arrived on the scene of the accident in an automobile and took the boy away with him to a hospital in Harlem. Jack Morgan left the park with his family, off to assume his destiny as the president of Morgan and Company. He was a responsible and dutiful son, hardly one of Upton Sinclair's rich wastrels, but his life would not be an entirely happy one. He would be forever haunted by the knowledge—his and others'—that he did not measure up to his father's imperious standards.

As for Terrance Tego, or Kego, whoever he was, he regained consciousness after arriving at the hospital, shocked and bruised but otherwise no worse for wear. He then promptly vanished from history. There is no record of what became of him in the twentieth century.

CHAPTER TWO

Thin Ice

JANUARY 6–JANUARY 31

> BEAVER FALLS, Penn., Jan 31.—Locked in each other's
> arms, Miss Zella Wylie, aged 19, daughter of the Rev. R.C.
> Wylie, pastor of the First Reformed Presbyterian Church of
> Wilkinsburg, Penn., and R.C. Patterson, aged 21, of Greens-
> burg, went down to death under the ice of the Beaver River this
> afternoon.
>
> Through the clear ice the bodies of Patterson and Miss Wylie
> could be seen at the bottom of the river.
>
> —The New York Times, February 1, 1908

Americans of 1908 were, in theory anyway, a prim people, constrained by rules of dress and decorum that would strike their
grandchildren and great-grandchildren as ludicrously restrictive. They
wore clothing that was heavy, hot, and impractical. Women suffered
spine-contorting corsets, covered by seven or eight layers of fabric, and
dragged around skirts that nearly scraped the ground. Men boxed themselves into square-shouldered three-piece suits, complete with waist-
coats they seldom removed, even in midsummer, and high stiff collars that
grasped their necks. For both sexes, hats were all but obligatory out of
doors, and women frequently draped veils from their hat brims to cover
their faces. Fresh air seldom touched human flesh.

Bound in their wardrobes, most American men and women entering
the twentieth century continued to observe Victorian rules of social
interaction—rules that were mainly meant to keep them apart. Recre-
ational sex was strictly prohibited. Women were expected to remain

chaste until their wedding night, and then suffer intercourse only to fulfill their sacred and patriotic obligation to populate the earth with more Americans. Men might enjoy the act—they could not help themselves—but were expected to feel ashamed of their more bestial natures afterward. Young mothers were harangued to be ever vigilant for evidence of masturbation in their children. One popular parenting guide provided an exhaustive catalogue of warning signs:

> Should you discover your child listless and preferring solitude rather than companionship, averse to exercise, averted looks, nervous, hypochondriacal, restless in sleep, constipated, pain in the back and lower extremities in the morning, appetite vacillating, hands cold and clammy; if you have already been suspicious, watch carefully now, even though not half these symptoms are present.

The New York State Insane Asylum, warned the book's author, was filled with habitual masturbators. A wonder it wasn't filled with hysterical young mothers.

Sex wasn't the only sensual pleasure that raised hackles. Moving pictures, that new entertainment to which the masses had taken with a passion starting around 1905, were condemned by ministers and concerned citizens as immoral and dangerous, appealing as they did to baser instincts. Liquor, too, was much maligned; half the states in the union had enacted temperance laws, most recently Georgia, which went completely dry on New Year's Day of 1908. Even Sunday comic supplements came under scrutiny from educational leaders who worried they might corrupt America's youth with undue silliness.

Altogether, Americans of 1908 believed they lived in an age in which humans could, and should, master the natural world, including their own appetites and instincts, by steady application of discipline, intelligence, and technology. Surely a civilization that could produce internal combustion engines, electric light, disposable razor blades, wireless telegraphy, and Kodak cameras was capable of fixing a little thing like human nature.

Had they succeeded, they would have made an admirable but tiresome generation. In fact, though, they mostly failed. For all their longing for order, and for all of their prosperity and technology, Americans of 1908 appeared, in certain lights, to be lurching out of control. They were a violent, giddy, sentimental people, flying off the handle at the slightest provocation. Suicide was epidemic, up from a rate of 1.2 per 100,000 in 1881 to 12.6 per 100,000 people by 1908. (Today, it's 10.6 per 100,000.) Anarchy was loose throughout the country. In New York City, the Black Hand dynamited tenements and storefronts. In western Kentucky and Tennessee, Night Riders galloped across the tobacco lands, lighting barns afire. Throughout the south, lynch mobs yanked black men out of prisons and hung them from trees. The divorce rate, another barometer of cultural upheaval, soared from one in every twenty-one marriages in 1890 to about one in every ten marriages by 1908. Amusement parks and dance halls roused couples into sweat-soaked deliriums. Young women walked the streets alone and took employment among men. Emma Goldman, the anarchist, spoke of a thing called "free love." Under all those prim clothes beat some primitive hearts.

This wasn't really news to Americans in 1908. Fifty years after the publication of Charles Darwin's *On the Origin of Species,* people grasped that humans evolved from lower animals, and were therefore prone to animal passions themselves. The theory of evolution, or "natural selection," was deeply embedded in their intellectual software. In fact, it was something of an obsession, one that lurked below every serious discussion of politics and culture.

Evolution meant different things to different people. Affluent whites, encouraged by social Darwinists like William Graham Sumner to interpret evolution as "survival of the fittest," looked to the theory as an endorsement of their own high status: they owed their prosperity not to luck or avarice, but to the laws of nature; they were rich and more cultured because they were more evolved, more "fit," than their more savage (and generally swarthier) lessers.

But Darwin's revelations also left many Americans anxiously scouring the culture for evidence of savagery or "degeneration," as the scientists called it. If it was no surprise that humans carried primitive impulses, it was nonetheless troubling to find that the modern world, with its fabulous new possibilities, did little to erase these. Progress evidently did not speed evolution. The primordial past was demonstrably not in retreat; on the contrary, it was recapitulating in new forms that were repellent to the codes of civilized morality, but also—and here was the truly disturbing part—profoundly alluring. With greater hopes of transcendence had come greater temptation for transgression.

*   *   *

As one possible symptom of concern that life was becoming dangerously
unshackled, newspapers were filled with cautionary tales of madness in
1908. Most metropolitan papers, including the *New York Times*, contained daily chronicles of unfortunates who had lost control and given
themselves over to deviant impulse and savage instinct, to unwhole-
someness and disorder. Formerly sensible people turned into raving
lunatics, screamed and thrashed, stuck their heads in ovens, jumped
from tall buildings, or climbed onto the Brooklyn Bridge and threw silver coins to newsboys on the deck below (as one man did a few days into
the New Year).

Perhaps it is only coincidental that newspapers that winter, especially
the *Times*, also devoted numerous articles to the perils of ice-skating.
Until a cold snap arrived in mid-February, not a week passed without the
Times running several tales of skating parties that had ventured out onto
the ice too early and suffered dreadful consequences. "Skating is a sport
in every way admirable and the eagerness for its exhilarating joys is as
comprehensible as it is innocent," a *Times* editorial noted early in the
month, "but life is too large a price to pay for them, and the perils of
reckless haste are enormous."

Ignoring caution—and the *Times*'s warnings—groups of young boys
or pairs of sweethearts glided out over a lake or pond in foolish delight.
Suddenly, the ice cracked and split. Somebody plunged through; rescue
attempts were made in vain; the victims succumbed to the numbing
water and drowned. The details of the stories were always a little different, but the moral was always the same. Beware. Beware the thin ice.

Thaw

The temperature dropped over the weekend, and the first Monday of the
New Year dawned at nineteen degrees in New York. Harry Thaw woke
up in his prison cell in the Tombs, blinking his protuberant eyes. Outside, through the thick stone walls, he could hear dray horses clacking
along the streets, the piercing wails of automobile horns, and newsboys
calling out the morning's news in the brittle frost: *Harry Thaw! Harry
Thaw!* The very name suggested thin ice.

The guards came for him late in the morning and escorted him across
the Bridge of Sighs, the elevated and enclosed walkway that joined the

Tombs to the Criminal Courts Building. Harry Thaw could look out through the small arched windows of the bridge to see a crowd gathered at the corner of Centre and Franklin Streets. The "familiar throng of loungers," as one of the papers described the crowd, had been standing there all morning in the cold, kept at a distance from the courthouse doors by police. Some were there no doubt because they were among the quarter of a million unemployed workers who had been wandering the city since the October crash and had no place better to go. Most, though, had come expressly to glimpse Harry Thaw, the most infamous madman in America.

Suddenly, there he was: a swoosh of dark fabric on the bridge, lit for a moment by a shaft of rising sun, then gone into the red-brick Criminal Courts Building of the Supreme Court of New York.

This day, January 6, marked the start of the second trial of Harry Thaw for the murder of Stanford White. The first trial, in the summer of 1907, had ended in a hung jury. Now the case was back in court for round two. Even before the start of this second trial, the case of Harry Thaw, Stanford White, and Evelyn Thaw had etched itself deep into the thoughts of early twentieth-century Americans. Anyone who read newspapers could recite its details in full, beginning with the warm summer night eighteen months earlier—June 25, 1906—when Harry and his pretty young wife entered the rooftop cabaret theater of Madison Square Garden. Stanford White was in the audience, sitting alone near the front, enjoying the performance of a light musical comedy, *Mamzelle Champagne.*

Stanford White, fifty-two years old, partner in the firm of McKim, Mead, and White, was one of the most celebrated architects in America. His work included the arch at Washington Square, a library at Columbia University, churches, municipal buildings, and countless mansions of the rich. But Madison Square Garden, with its exotic Moorish details and soaring tower, was his masterpiece and, in some senses, his own true home. He kept an apartment high in the tower, a pied-à-terre for those frequent evenings he failed to make it home to his wife and family on Long Island. It was here on Madison Square that his professional and extracurricular interests came together—and where his life, with fine dramatic closure, came to an end. He was listening to a soloist sing "I Could Love a Million Girls"—a fitting requiem, under the circumstances—when Harry Thaw stepped up behind him and pulled out a revolver.

Until this moment, Harry Kendall Thaw had distinguished himself mainly as a profligate and petulant young man. His late father, William Thaw, had made a fortune in the steel and coke industries in Pittsburgh, and Harry had done his best to squander it. He'd briefly attended Harvard College but had been expelled. Since then, he'd devoted himself to pursuing women, playing cards, and turning over tables in restaurants. According to one tale, he'd once spent eight thousand dollars on a dinner in Paris. On another occasion, he'd attempted to ride a horse up the stairs of New York's Union League, enraged that the club had refused him admittance.

Still, none of this prepared anyone for what he did next: he set the muzzle of his revolver inches from Stanford White's head and pulled the trigger three times. As White reeled in his seat, then slumped over, Harry Thaw stepped back, raised his arms, as if to signal that his work was done, then walked back, amid ensuing pandemonium, to join Evelyn.

"Good God, Harry," she cried out. "What have you done?"

"It's all right, dear," he assured her. "I have saved your life."

The public murder of one of the city's grandest personages by one of its richest instantly riveted New York. But it was the revelations that surfaced in the deluge of press coverage after the murder that carried the most remarkable news. To begin, Stanford White was not quite the paragon of respectability for which better society took him. He had a predilection for throwing stag parties and deflowering adolescent girls. This explained the special little mirror-lined apartment he'd fitted for himself in Madison Square Garden. It also explained his attraction, in 1901, to Evelyn Nesbit, then a sweet-faced sixteen-year-old chorus girl from Pittsburgh.

The highlight of Harry Thaw's first trial had been Evelyn's testimony regarding the particulars of her relationship with White. Wearing a schoolgirl outfit and looking not much older than she'd been when she met White six years earlier, she had taken the stand and tearfully recounted her deflowering in detail: how White brought her late one evening to his little apartment in Madison Square Garden and there, amid the mirrors and imported rugs, pushed her on a red velvet swing that hung from the rafters; how he served her champagne, begged her to change into a yellow silk kimono, then trembled at the sight of her when she did. She passed out, and when she came to she felt a pain between her legs and saw blood on the sheets. White stood over her, gloating in triumph.

No woman had ever publicly discussed sex like this, much less a pretty young woman like Evelyn, and the effect was riveting to those who read the front-page trial transcripts in newspapers. For the press, the publication of Evelyn's faithfully recorded testimony marked a journalistic milestone of sorts, not one of which every editor was proud. "The worst thing about the first Thaw trial was that there was no verdict: that the odious story needs be retold," grumbled the *Los Angeles Times* at the start of the second trial. "Few there be, of sane, normal men and women who would care to again hear or read Evelyn Thaw's story of her degradation."

The *Los Angeles Times* nonetheless reported on the second trial about as diligently as every other newspaper in the country. No publisher or editor who valued his readers would have dared to do otherwise. At a time long before radio and television allowed millions to share experience simultaneously by simply turning a dial, the Thaw case was a rare national event, one that reached across state lines into nearly every household, bruited on the front pages of newspapers like the *Dallas Morning News,* the *Grand Forks Herald,* the *Macon Weekly Telegraph,* and hundreds of others throughout the country.

So many elements combined to make the case fascinating to American readers. The frank talk of sex, of course, but also the fact that it took place among the upper echelons of society and portrayed them in a rare and shadowy light. "The flash of that pistol lighted up an abyss of moral turpitude, revealing powerful, reckless, openly flaunted wealth," is how William Randolph Hearst's populist *Evening Journal* put it. Both Harry Thaw and Stanford White were well-heeled, well-bred Victorian gentlemen, yet each had behaved like an animal. Out of all the reportage and editorializing emerged an important if not exactly shocking news flash: the supposedly evolved rich were as primitive in their morals and appetites as the uncivilized poor.

Just after 11:30 A.M. on that January morning, Judge Victor Dowling called the court to order. The door opened and Harry Thaw entered the small overstuffed room, walking as usual with his left shoulder raised slightly higher than his right. The reporters studied him, trying to assess his mood, his health, his state of mind. To the reporter from the *New York Herald,* Thaw's pale skin, offset by unruly black hair, cast an "ashen hue," while his bulging eyes held the glare "of a haunted man." Most of the reporters, though, agreed that Thaw looked more fit and calm this morning than at his previous trial. Incarceration seemed to agree with him. The reporters knew he had been exercising regularly with a medicine ball in the prison yard of the Tombs and had recently become a devotee of Christian Science. They knew, too, that he had been living in relative luxury, having bribed jail officials to cart in a few amenities to accommodate his tastes, including Tiffany lamps and meals ordered from Delmonico's.

Once Thaw was seated, the attention of reporters turned to the more compelling subject of his wife. Evelyn was perched in the spectators' gallery about ten feet behind Thaw. Her small hands clasping the railing, she stared ahead "with the same wistful intent gaze" she had exhibited at the first trial. She wore the same blue schoolgirl outfit, too, with the short skirt that rose almost to the tops of her shoes. Her black velvet hat was plumed with a spray of violets.

The wardrobe had not changed much since the last trial but the woman had. She was more grown-up, more self-possessed. "Time seems to have made a difference," she told a reporter from the *Chicago Tribune*. "Last year, when the newspapers would say mean things about me I used to cry and feel bad. I like to read the accounts in the papers, and now I don't seem to mind so much when they say mean things." Several newspapers approvingly noted a small weight gain on her petite frame that made her oval face appear softer, her cheeks more flushed, her figure fuller and more womanly. She told reporters that she was eating well and taking good care of herself this time around. "I take a cold bath every morning, about as cold as I can get it, and send the cold water down my spine the first thing. Then I drink sterilized milk."

With her soft cheeks, her plush waves of hair, her wide eyes and smile—not to mention her innocent but stirring talk of cold baths and warm milk—Evelyn put a sweet face on a sordid tale. Either by instinct or with more artfulness than she let on, she played the role of America's first truly national sex symbol with perfect pitch, and Americans responded devotedly. Newspaper readers could recount when and under what circumstances she surrendered her virginity to Stanford White. They could speak of Harry's tendency to beat her, and of how, once, in a castle in Germany, he tied her to a bedpost and whipped her. They knew she had been ill-used by her own mother, who passively consented to her concubinage by White; had been ill-used, in fact, by just about everyone who ever looked into her pretty winsome face and saw sexual bliss or financial profit smiling back. And they knew, finally, how she had been forced to publicly recount every one of her degradations during the first trial of Harry Thaw. And yet, for all this—or, rather, because of it—she was more luscious and radiant than ever.

Evelyn was one of a small number of women in the courtroom this morning. The judge had barred women spectators, allowing only female newspaper reporters and family members to enter. A few of the latter were absent. Harry's mother was home in Pittsburgh, ill in her bed. His favorite sister, the countess of Yarmouth, was detained in London, attempting to annul her seven-year marriage to the "effeminate" earl of Yarmouth on the grounds that it had never been consummated—one more international marriage gone bad. Another sister, Mrs. George Lauder Carnegie, wife of Andrew Carnegie's nephew, was present, suffering a bad head cold and barely willing to acknowledge Evelyn.

The trial began with sleepy preliminaries. As the radiator hissed and clicked, and Mrs. George Lauder Carnegie coughed and sniffled, the attorney began interviewing talesmen. When something amusing was uttered, Evelyn Thaw was seen to smile behind a gloved hand. Now and then, her husband turned and mouthed some words to her, and she mouthed back, as if she understood perfectly what he was trying to say.

Evelyn Nesbit, around the time Stanford White met her in 1901. Photograph by Rudolph Eickemeyer Jr.

Crossing the Line

The second trial of Harry Thaw was not the only news commanding
national attention in those early days of January. Another drama, far
more epic in scope, albeit less scintillating in content, was unfolding that
same Monday morning thirty-two hundred miles from New York City.
In the southern Atlantic Ocean, due east of the mouth of the Amazon
River, sixteen American battleships, painted white and billowing thick
black smoke, were steaming southwest on a voyage around the world.

The progress of the Great White Fleet, as this American armada
would come to be called, was reported in newspapers side by side with
stories about the Thaw trial and provided a purifying contrast to that
seamy spectacle. Here was American manhood at its worthiest: thirteen
thousand sailors aboard sixteen battleships, the men crisp in navy blues,
the ships gleaming white, all plowing through the ocean in perfect for-
mation at exactly ten knots. As fodder for journalism, the Great White
Fleet lacked Evelyn Thaw's sexual divulgences and Harry Thaw's eye-
popping strangeness, but as a demonstration of wholesome American
virility, nothing surpassed it.

By coincidence, the men aboard the battleships were undergoing their
own trial that first Monday of the year, one arguably more unpleasant than
anything Harry Thaw was enduring on the opening day of his trial on
Centre Street. The previous evening, shortly before midnight on January
5, 1908, the sixteen battleships had passed over the equator—"crossing the
line," in navy jargon. By long-standing naval tradition, any sailor aboard
a U.S. vessel who had not previously crossed the equator was brought
before the so-called Court of Neptune (a few old salts dressed up in
absurd costumes) and condemned to an elaborate hazing ceremony.
Since the neophytes aboard the American battleships on this January
morning constituted roughly 90 percent of the thirteen thousand men of
the fleet, the hazing went on for most of the day. Men were slathered in
molasses and cylinder oil, sprinkled with flour, force-fed soap and oil, then
tipped backward into a tarpaulin filled with seawater—less and less
water as the day went on and the tarpaulin sprung leaks. None of this
could have been much fun to the men, but at least the hazing was a diversion from the monotony that had already begun to settle over the routine

on the ships, just three weeks into a voyage that would end up lasting fourteen months.

The voyage of the Great White Fleet had commenced with an elaborate send-off on the bright windless morning of December 16, 1907, in Hampton Roads, Virginia. Thousands of spectators had turned out along the shores to watch it depart. Earlier in the morning, President Theodore Roosevelt had steamed in aboard the presidential yacht, the *Mayflower*, to give a few last-minute instructions to fleet commanders and to add his own heft to the pomp and circumstance. As sailors in dress uniform stood at the rails and brass bands played on the vessels, the president watched with glee. "Did you ever see such a fleet and such a day?" he shouted to his guests aboard the *Mayflower*. "Isn't it magnificent? Oughtn't we all to feel proud?" It was all, concluded the president with his all-purpose exclamatory, "perfectly bully."

For sheer majesty, the Great White Fleet was as impressive a display of naval clout as America, or any other nation for that matter, had ever produced. "The greatest fleet of war vessels ever assembled under one flag," is how the *New York Times* described it. *Harper's Weekly* ranked it as more impressive than the Spanish Armada. The figures to back this claim were considerable. The sixteen vessels of the fleet comprised nearly $100 million worth of battleship and nearly a quarter million tons of armament. To fire a single broadside from all the ships' guns would cost roughly $50,000, Teddy Roosevelt's annual salary as president. In addition to their combined crews of 12,793 officers and men, the ships carried 1.2 million pounds of flour, 100 million pounds of beef, 800,000 pounds of potatoes, 1 million pounds of vegetables, 12,000 pounds of pickles, 30,000 pounds of sausage, and 8,000 pounds of jam. Also aboard were twenty-five goats, thirty-two dogs, twelve parrots, one donkey, and a sizable corps of journalists assigned to travel with the fleet on behalf of America's magazines and newspapers.

Thus loaded to the gunnels and pulling deep drafts, the ships steamed away, stretching out into a three-mile column. The *Mayflower* led them to the mouth of the Chesapeake Bay, and there, as the ships' bands played "The Girl I Left Behind Me," President Roosevelt gave a last wave of his top hat.

To those who witnessed its departure that day, and to the many Americans who read about it in the weeks and months after, the voyage

President Roosevelt bids farewell to a few of the Great White Fleet's
thirteen thousand sailors as they begin their forty-three-thousand-mile
journey around the world on December 16, 1907.

of the fleet told more about America's new place in the world than any
statistics of economics or demographics. Ten years earlier, the Spanish-
American war had heralded the nation's debut as a worthy player on the
world stage. In the decade since, the country's economy had expanded
vastly and the navy had grown commensurately. America now possessed
a navy second only to Great Britain's and on a par with Germany's. And
no one doubted that if the need should ever arise, America, with its
capacity to produce more steel than Britain and Germany combined,
could build ships faster than any country on earth.

If there was little doubt of the fleet's might, there remained the more troubling question of its mission. Where exactly were those sixteen battleships
going—and why? Officially, the voyage was to end in San Francisco
after rounding the bottom of South America. But by the time the fleet left
Hampton Roads, it was an open secret that President Roosevelt intended
to send the ships all the way around the world. The exact itinerary had not
been worked out yet; most likely the fleet would pass through the Suez

Canal before returning to Hampton Roads. Again, all this begged the question: why?

The question is surprisingly difficult to answer even now. At the time, no less an authority than Rear Admiral Robley "Fighting Bob" Evans, the commander in chief of the fleet, claimed ignorance of the fleet's ultimate purpose. The subject was simply never broached in his prevoyage discussions with the president. "There was no reason why it should have been," Evans later wrote. "My business was merely to obey his orders when I received them and see that the fleet was properly handled, which I did."

This much was certain: the decision to send the Atlantic fleet into the Pacific belonged exclusively to President Roosevelt. He later admitted—boasted, rather—that he made it without consulting his cabinet, "precisely as I took Panama without consulting the cabinet." To have done otherwise, he explained in his autobiography, would have been "to take refuge behind the generally timid wisdom of a multitude of councilors."

A former assistant secretary of the navy and author of a book of naval history, Roosevelt had grounds to consider himself qualified to make decisions regarding the deployment of America's fleet. Though he never fully explained his decision to send the fleet on this worldwide cruise, he did give several plausible reasons, both at the time of the fleet's departure and in later writings. Most frequently, and least convincingly, he characterized the voyage as an opportunity for the navy to sharpen its skills and tactics; an elaborate training exercise, in other words. This was a fine enough explanation up to a point, but it hardly seemed to justify a forty-three-thousand-mile voyage around the world. Would not eight or ten thousand miles have sufficed?

Roosevelt also made clear that he intended the voyage to show off America's naval power. Not because he was vain about it—though undoubtedly he was—but because he thought the world (as well as his fellow Americans) ought to appreciate how mighty the U.S. Navy had become. Roosevelt's diplomatic policy, as he never tired of reminding people, was "speak softly and carry a big stick." At a moment when Britain had recently sparked an arms race with Germany by unveiling an intimidating new supership, HMS *Dreadnought*, and the Germans had responded in kind, the Great White Fleet was a very big stick meant to remind friend and foe alike that America was not to be taken lightly in matters of global concern. What better way to exhibit America's big stick than to sail it around the world and wag it for all to see?

Stating this in the sort of stark Darwinian terms that would have been familiar to readers in 1908, one contemporary writer promoted the fleet's voyage as a necessary demonstration of American muscle in a world of potential antagonists. "It is well for us to listen to beautiful sermons on the continual peace of the world," Robert D. Jones wrote in his forward to a 1908 book about the fleet, but "if the trend of things is to be resolved into a condition of 'the survival of the fittest' let us accept this condition and do our utmost to put ourselves in the place of most fit."

In addition to these broadly defined goals, there was a very specific objective of the fleet's mission, never articulated outright by Roosevelt but insinuated at every turn. This was to pointedly warn one potential antagonist, the Imperial Government of Japan, to avoid tangling with the United States of America.

In late 1907 and early 1908, America stood closer to war with Japan than it had ever been—or would be again for decades. The origins of these tensions were many, but their primary cause was simple racial bigotry on the part of xenophobic Americans. For several years, Japanese immigrants, along with others of Asian heritage, had been routinely segregated and discriminated against by Californians who believed the foreigners were taking jobs from them. The Japanese government and press understandably took offense at the treatment of its citizens in America. In response, some U.S. newspapers, especially those of the yellow press, had begun to write of armed conflict with Japan as inevitable.

Looking back, a local racial skirmish seems a flimsy pretext for war. Indeed, it seemed so to many people at the time, and the Roosevelt administration took pains to dismiss the prospect of a military engagement with Japan as far-fetched. But these dismissals were only partly sincere. Roosevelt harbored real concerns that Japan, newly emboldened by a recent naval victory over Russia, might in fact pose a threat to the Philippines and other U.S. interests. He wanted the Japanese to understand in no uncertain terms that America was prepared to fight. "I had been doing my best to be polite to the Japanese and had finally become uncomfortably conscious of a very, very slight undertone of veiled truculence," he would write a few years later of his decision to send out the fleet. "[I]t was time for a show down."

Of course, in attempting to impress upon Japan the folly of war with the

U.S., Roosevelt ran the risk he might provoke a war. From the moment the fleet left Hampton Roads, rumors began sweeping through diplomatic and journalistic circles that he had done just that. Nearly every day fresh bulletins of sinister Japanese maneuvers appeared in the European and American press: ten thousand Japanese had been discovered hiding out in Mexico, disguised as Mexican peasants and preparing to attack America; the Japanese navy had been spotted off the coast of Hawaii; a flotilla of Japanese torpedo boats was lying in wait in the Strait of Magellan, ready to pounce when the fleet arrived there in February.

The threat of war did not seem to trouble the president much. If Japan took this opportunity to strike, he later wrote, deploying a bit of circular logic, it was "proof positive that we were going to be attacked anyhow." All the better, then, to have the fleet out on the water, ready to fight.

Meanwhile, the voyage raised a few other concerns. Removing the entire Atlantic fleet to the Pacific would leave the Atlantic seaboard vulnerable to naval attack. And to get to the Pacific, the fleet would have to navigate the Strait of Magellan. Even putting aside the Japanese warships rumored to be lurking there, the rock-infested and tidally capricious strait was the most infamously treacherous marine channel in the world. Which raised the specter of $100 million of American battleships dashed to shreds before they got a chance to fight.

Worries about the voyage dovetailed back to the same nagging question: was President Roosevelt's decision to send the fleet to the Pacific a smartly considered act of realpolitik or was it a rash, costly, and potentially disastrous show of saber rattling? Some Americans were convinced it was the latter. The chairman of the Senate Naval Affairs Committee, Senator Eugene Hale of Maine, so opposed the plan he threatened to block congressional funding for it. This turned out to be an idle threat, since the president had enough money at his disposal to send the fleet to the Pacific without congressional approval. If the Senate wished to refuse funding to get it back to American shores, well, that would be their problem.

Roosevelt's bravado did little to assure those who already considered him a half-cocked warmonger. The hero of San Juan Hill had never made a secret of his belief that an occasional war was invigorating medicine for a country. Nor did it seem implausible to these same concerned citizens

that Roosevelt would risk a major naval battle to satisfy his own personal
vanity. Of course, only a madman would do such a thing. Then again, in
the words of Mark Twain, the president was "clearly insane . . . and
insanest upon war and its supreme glories."

Mark Twain often spoke with his tongue in his cheek, but in this case
his assessment was shared by far less humorous Americans. Rumors of
psychological illness had dogged Roosevelt almost from the day he took
office in 1901, following the assassination of President McKinley, and
began to conduct himself as no president in American history had ever
done. Within weeks of moving into the White House, he'd shocked the
world by inviting Booker T. Washington—an African American—to
dine with him and his wife. This racial faux pas inflamed much of the
south, where newspapers branded him "a rank negrophilist" (among the
more polite epithets), and upset a good part of the north as well. A few
months later, he'd incited Wall Street and broken a fifty-year tradition of
laissez-faire government by prosecuting J. P. Morgan's Northern Securi-
ties Corporation under the Sherman Antitrust Act, thereby attacking
one of the largest business interests in the country.

In retrospect, Roosevelt's actions appear courageous and forward-
thinking; at the time, they caused some Americans, including his political supporters, to seriously question his judgment. It didn't help that
Roosevelt tended to combine bold moves with an unpredictability that
struck others as, if not mad, then at least maddening. He dashed "from
one thing to another with incredible dispatch," as Twain put it, "each act
of his, and each opinion expressed . . . likely to abolish or controvert
some previous act or expressed opinion." One moment, the president
was excoriating big business for its excesses, the next upbraiding labor
unions for their abuses; one moment, calling for fair treatment of blacks,
the next unfairly discharging an entire regiment of black soldiers for the
misbehavior of a few; one moment, Bible-thumping like an evangelical
minister, the next removing the words "In God We Trust" from American currency. He had thus managed during his seven years in office to
alienate southern Democrats, African Americans, organized labor, Wall
Street bankers, and fundamentalist Christians, and yet remain, as Twain
put it, "the most popular human being that has ever existed in the
United States."

The truth was that no one could fail to be dazzled by a man so bursting with energy and opinions, however inconsistent they might be. For

better and worse, Roosevelt was the very embodiment of American speed, vitality, youth, and confidence. He spoke exhaustively on nearly every subject, with a know-it-all mastery that would have been annoying had it not been so impressive. In his pursuit of the "strenuous life," as he called it, he took exhausting, life-threatening hikes through Rock Creek Park and stripped for nude swims in the icy Potomac. Inside the White House, he engaged in raucous pillow fights with his children, practiced jujitsu, and boxed in the basement gym so ferociously that an opponent accidentally blinded him in one eye. With his other eye, he continued to devour two or three books a night, write thousands of words a day, and shoot wild animals every chance he got. How to explain such a man? Many of Roosevelt's own contemporaries dismissed him as a drunk, despite the fact he imbibed alcohol only sparingly. Henry Adams was probably closer to the truth when he described the president as "drunk on himself."

The crews onboard the battleships had no need to explain Roosevelt as they rounded the eastern hump of South America. They were simply following orders, too removed in their day-to-day concerns to wonder whether this venture would end in tragedy or triumph. Either of those two outcomes was preferable, in any case, to the comedy of errors unfolding under the equatorial sky in the early days of 1908.

The first element of farce appeared in an article in the January issue of *McClure's* magazine. Entitled "The Needs of the Navy," the article was written by Henry Reuterdahl, a highly regarded naval expert (who happened to be accompanying the fleet aboard the battleship *Minnesota*). Reuterdahl's article argued that the battleships of the American fleet were riddled with design flaws, the most significant being the position of the armor belt that encircled their hulls. In no battleship did this reach more than six inches above the waterline. When the ships were weighted with coal and supplies, as they were when they left Hampton Roads, the armor belts on some of them were actually *underwater*, making their flanks entirely vulnerable to attack. For all their fierce armaments, the ships of the Great White Fleet were sitting ducks.

Another potential embarrassment, minor but troubling, came to light around the time the fleet was passing over the equator. Admiral Evans discovered that somebody on his staff—no names mentioned—had miscalculated the distance between Trinidad, their first port of call, and

Rio, their second. According to the prevoyage specifications, the distance was 2,900 nautical miles. In fact, the distance was more like 3,300 nautical miles. "Somebody," the admiral admitted, "must have blundered egregiously." The extra miles would delay the fleet's arrival in Rio. More worrisome was the possibility that a few of the larger vessels would run out of coal before they ever got there, leaving the pride of America adrift somewhere off the coast of Brazil.

For now, though, life with the fleet passed in the monotony of shipboard chores. Leisure consisted of dining in the messes, watching movies, and, on warm nights when the band played on the deck, waltzing in each other's arms. As they danced, they dreamed of sweethearts back home; or, if they had none, perhaps they conjured up Evelyn Thaw, as military men of the future would moon over, say, Rita Hayworth or Marilyn Monroe. Ideally, the Great White Fleet was an invulnerable armada of disciplined and wholesome sailors. In reality, it was sixteen mortal vessels running behind schedule and populated by thousands of horny young men.

"It is a beautiful sight to look down the long lines and see the signal lights flashing and the hulls shining in the moonlight," wrote Henry Davis, a captain aboard the battleship *Ohio*. "If only some arrangement might be made by which we could occasionally see and talk to some of the fair sex, life would not be so stupid."

The Great White Fleet at sea

The Unwritten Law

On Monday, January 13, the Thaw trial began in earnest. Jury selection had consumed an entire week, days and nights, but finally, after interviewing 372 talesmen, the attorneys were settled on 12 jurors and ready to deliver opening statements.

Thaw's chief attorney, Warren Littleton, began with an overview of the insanity defense he intended to present. "I tell you the story of the life of Harry Thaw to the end," he began, "that you may see, as with the eyes of his mother, his teacher, his family, the manner of the man. I bring you no story of a man who was strong and virile and active, and who suddenly, under the power and acuteness of a passion, was overthrown, and then suddenly restored again." Rather, he assured the jury, this would be the story of a man insane throughout his strange, wasted life.

Insanity had been the plea in the first trial, too, but it had been a more proscribed form then. Thaw's attorney at that trial, Delphin Delmas, had claimed that at the moment Thaw shot White, he was suffering from a condition Delmas coined as "*dementia Americana*." The attorney defined this as the rage that any red-blooded American male was likely to experience after discovering that the sanctity of his home and virtue of his wife had been violated: "that species of insanity that inspires every American to believe his home is sacred." By killing White, Delmas argued, his client had followed an "unwritten law" and "struck out for the purity of wives and the homes of America."

If the *dementia Americana* argument was specious, it was also ingenious, invoking as it did a principle most Americans took for granted—that women were fragile and vulnerable creatures who depended on the protection of men. It was on this principle that Judge Dowling had barred women spectators from the Thaw trial; best not to expose them to the brute facts of the case. More significantly—and surprisingly, from a twenty-first-century point of view—it was on this principle that a brilliant young progressive attorney named Louis Brandeis was about to frame a landmark argument before the U.S. Supreme Court. Appearing before that court on Wednesday, January 15, in the case of *Muller v. Oregon*, Brandeis would defend an Oregon statute that prohibited women from working more than ten hours a day in physically demanding occupations. Women, he would argue (providing reams of sociological data to support his case), were a special class of citizen who required and deserved special protections under the constitution of the United States. The Supreme Court would unanimously agree.

The idea that women needed men and government to protect them, either with a gun (like Harry Thaw's) or a law (like Oregon's), was more a product of wishful thinking than an accurate reflection of reality in 1908. More than a fifth of American women worked outside the home, with as many as a third working in cities. These working women hardly lived protected lives, and many hardly wanted protection. But protection they would have. In some cases, this patronizing attitude led to arguably positive outcomes, as in *Muller v. Oregon*. More often, it gave rise to fatuous twaddle like New York City's short-lived Sullivan Act, passed in late January of 1908 to prohibit women from smoking in public, as they had done in numbers on New Year's Eve. (The act prompted a visiting French journalist to remark to the *New York Times*, "There is no place where it is easier for a woman to earn a livelihood than in the United States, nor where it is more difficult to smoke a cigarette.")

Putting aside larger social realities, Delmas's interpretation of the murder required jurors to gloss over a few inconvenient facts. Such as, for instance, that Evelyn's affair with White had occurred several years before she and Harry married, so the wifely virtue Thaw was supposedly protecting was neither wifely at the time nor especially virtuous. Trickier than painting Evelyn as an innocent maiden was portraying Harry, a spoiled alcoholic who liked to flog women, as a chivalric hero. Whatever its flaws, though, Delmas's *dementia Americana* argument had nearly prevailed in the first trial of Harry Thaw. The jury hung, and Thaw did not.

For this second trial, Warren Littleton was having none of his predecessor's Victorian morality play. He settled, instead, on the tale of a madman. Harry Thaw, as Littleton told the jury, was truly insane—chronically, constitutionally insane—from birth.

"Let us look back and understand what manner of man this was at the beginning, for no man can escape the despotism of that destiny, that which was decreed before he was born; no man can rebel against that tyranny which was established over him by his ancestors; no man can flee that which burns in his blood at birth, and will burn in his blood until he dies." Littleton's argument relied on another popular assumption in that Darwin-besotted age. Psychologists—or alienists, as they were often called at the time—believed that madness was a hereditary flaw, a "taint" that had little to do with life experience and everything to do with bloodlines. Harry's bloodlines, Littleton would have the jury believe, were heavily tainted.

The argument was an uncomfortable one for the Thaws, since it required the attorney to more or less shake the family tree and see what nuts fell loose. Littleton ran through a long list of mad or melancholic ancestors to establish a history of lunacy in the Thaw family. Then he called an array of witnesses to attest to Harry's lifelong oddities. Harry's childhood physician told the jury that he found the boy to be highly nervous and of "unsound mind." Harry's old teacher recalled how little Harry would sometimes tear at his clothes and throw chairs against the wall. Another teacher recalled Thaw's "sudden, passionate outbursts, his animal-like howls," and described his appearance as that of an "unwholesome boy," a euphemism that called up unsavory images of young Harry masturbating his way to madness. Several alienists came to the stand to apply a diagnosis to Harry that few Americans had ever heard before: *manic-depressive.*

On Friday, January 17, Harry's mother, Mrs. William Thaw, arrived at court, despite a prolonged illness. Her health was improved enough for her to make the trip from Pittsburgh, but she looked aged, feeble, and altogether pitiful in the eyes of the press. This trial must have been an especially difficult moment for this lady of highest Pittsburgh society. Obliging Littleton, she told the jury how Harry had screamed and twitched as an infant and later suffered from St. Vitus's dance; how he had nearly driven her to despair with his sleeplessness and tantrums and other oddities. By the time she was finished, she'd done a convincing job of portraying her son as a born loon.

As Mrs. Thaw stepped wearily off the witness stand, she was replaced by the star attraction, her daughter-in-law, Evelyn, and the mood of the courtroom instantly lightened. Journalists could not help noticing the striking contrast between mother and wife. "In the physical appearance of added bloom, rounded contour and abounding health," observed the *Herald,* "the year that has elapsed seemed to have done as much for the improvement of the chorus girl whom Harry Thaw made his wife as it had done in unsparing detriment to the woman who first loved him."

According to press reports, Evelyn's high spirits had been purchased by a large sum paid to her by the Thaw family in exchange for testimony favorable to Harry. Perhaps, too, her mood owed something to a certain bon vivant with whom she'd been seen dining about town late at night. Whatever the truth of these rumors, she appeared the very soul of comely virtue as she reprised her testimony of the previous trial, using almost the same words she'd used a year earlier to describe events leading up to the murder of Stanford White. Some of her same girlish mannerisms were on display, too. The pretty pout when the DA rose to object to something she'd said; the gloved finger coquettishly brushing her lips; the helpless glances at the judge.

Really, though, there was nothing helpless about Evelyn this time around. She spoke quickly, matter-of-factly recounting her impoverished youth near Pittsburgh, her arrival in New York at the age of fifteen to work as a model and chorus girl, her introduction to, and eventual seduction by, Stanford White. In the first trial, she'd struggled through this last part with tears and blushes. This time, she told it with little effort and less drama. "The repetition lacked the vitality of the first recital," noted the same *Los Angeles Times* that had earlier decried the trial for its unsavory content, sounding almost disappointed now.

In his cross-examination, the district attorney, William Travers Jerome, seemed intent on discrediting Evelyn by establishing her as a promiscuous floozy. He sneeringly inquired whether she'd continued to sleep with Stanford White even after she'd become Thaw's mistress, an interesting line of questioning in its own right, perhaps, but not salient to the sanity or insanity of Harry Thaw. Nor did Evelyn allow the district attorney's insinuations to pass unchallenged. "Despite her extremely youthful appearance," reported the *Chicago Daily Tribune*, "she soon became, under Mr. Jerome's merciless fire, an extremely clever, resourceful, self-contained young woman, who saw just where any given question led almost before it had left the questioner's lips."

Before ending his cross-examination, Mr. Jerome did pry from Evelyn the potentially significant admission that Thaw had seemed perfectly rational to her in the hours before he killed White, but not even the district attorney seemed to care much about proving the defendant sane anymore. He ended the trial spraying scorn on everyone involved, from Stanford White to the alienists who'd testified in Thaw's defense, and most of all on Thaw himself; or, as Jerome referred to the defendant, "This miserable creature, this pervert."

Nearly a month had passed. Some of the steam had seeped out of that overheated courtroom. The first time through, the spectacle of a young woman responding frankly to an interrogation about her sex life had been shocking and titillating; this time, it was merely depressing. Something had changed between that first trial and this one. The first trial itself probably had a lot to do with this. As any halfway sentient newspaper reader understood by late January of 1908, this was not a quaint melodrama of good avenging evil. Casting Harry Thaw in the role of young hero upholding the codes of civilized morality might have been appealing, conforming as it did to Victorian conventions, but it was also blatantly false. If this man was the standard-bearer of Victorian values, those values were corrupt to the core.

The disturbing possibility suggested by the Thaw trial was that Thaw and White really *were* standard-bearers of Victorianism; that both men were the express products of a morality that pretended women were sexually innocent, if not entirely numb, and men were their protectors; that Victorianism made a fetish of female innocence and created men like White, who worshiped female virginity so he could take it, and Thaw, who beat women mercilessly for not holding on to it.

It would be facile to blame Thaw's murder of White on Victorian repression; more facile still to suggest that Americans all at once tossed aside Victorianism in the wake of the Thaw trial and embraced modern views of sexuality. They did not. But merely by its placement on the front pages of their newspapers, the Thaw trial held the old morality up for daily inspection and demanded, at the very least, an acknowledgment of its ironies and hypocrisies.

A foot of snow blanketed the city the last weekend of January. The flakes began to fall on Thursday, then continued into Friday, and by Saturday morning the city was still and quiet. The laborers hired by the city to remove the snow at twenty cents per cubic yard went on strike, and huge banks of it piled up on curbs and corners. Streetcar service halted. Automobiles were immobilized, their cacophony silenced. The most useful form of transportation that weekend was the horse-drawn sleigh, and sleighing parties swished through the parks, bells jingling. The city seemed to lapse back into an earlier, quieter time in its history. But then the shovels came out and it was back to the usual frantic twentieth-century rhythms.

Shortly before noon on Friday morning, January 31, the jury in the case of the State of New York against Harry Thaw went into deliberations to determine whether the defendant was guilty of murder and therefore headed for the electric chair, or merely insane and destined for the asylum.

CHAPTER THREE

Mysteries of the Unknown Parts

JANUARY 31–FEBRUARY 12

Never in the history of modern exploration have efforts so
widespread and persistent as those of the present been made to
uncover the mysteries of the unknown parts of the world.
—*The New York Times*, February 2, 1908

Early on the afternoon of January 31, just as the Thaw jury was beginning its deliberations in New York, the battleships of the Great White Fleet rounded the southern tip of Argentina and turned westward into the mouth of the Strait of Magellan. The fog lifted and a low and treeless landscape revealed itself in the distance. "South of us we can just see Tierra del Fuego," Louis Maxfield, a midshipman, wrote home from aboard the *Illinois*. "A mile north is the most barren and desolate landscape you can imagine."

The channel they were about to navigate ran more than three hundred turbulent miles from the Atlantic Ocean to the Pacific Ocean. It was as narrow as two miles and teemed with unseen shoals and jagged outcroppings of rock. No modern fleet of warships had ever attempted to navigate it.

"This will be in reality the most momentous point in the entire journey," the *New York Tribune* prophesied. "[I]f trouble has any favorite abiding place in all the seven seas it is right in these same deep, narrow and storm racked waters."

For shipboard journalists who been dulling their pencils on hazing ceremonies and garden parties (of which there had been many during

55

the stop in Rio), here at last, amid these forbidding, primeval surroundings, was an opportunity to write spine-tingling prose. They made the
most of it in the stories they sent home by way of the ships' newly
installed wireless telegraph systems. This "bleak outpost of the world,"
the *New York Herald* informed its readers, was a "savage waterway"
enclosed by "peaks of naked rock shooting vertically out of the spinning
tide for two or three thousand feet, uncouth, repellant, and sublime." To
which the *Tribune* added that "dangerous rocks abound and swift currents sweep with merciless energy."

"We seem to be," Midshipman Louis Maxfield wrote simply, "at the
end of the world."

That same day, at another end of the world, the man responsible for dis-
patching Louis Maxfield and his fellow sailors to the Strait of Magellan
was once again stirring up concerns that his mental apparatus was nearly
as flawed as poor Harry Thaw's.

Theodore Roosevelt chose this day to send down to both houses of
Congress a "Special Message," one of his regular communiqués to the leg-
islative branch. He had sent a good many of these Special Messages to
Congress during his years in office, and most had been verbose and
strongly worded. Never, though, had he sent a message quite like this one.
It was, as the *Times* described it in a broad headline, the HOTTEST MES-
SAGE EVER SENT TO CONGRESS.

The main body of the message was a harsh indictment of "certain
wealthy men," as Roosevelt put it, "whose conduct should be abhorrent
to every man of ordinarily decent conscience." Loading on fire-and-
brimstone adjectives—"evil," "sinister," "wicked"—Roosevelt charged
the immoral rich with oppressing wage workers, defrauding the public
with stock schemes, and generally sacrificing the common good for their
own greed. Roosevelt's intention to hold the rich accountable for their
sins was hardly news. What was different here was not the substance, but
the unrestrained fury of the language—the "sledgehammer eloquence,"
as the *Boston Globe* described it.

For the president's admirers, the message was a resolute call for
morality and accountability among the rich. To his detractors, it was further evidence the man was seriously unhinged. "In the unseemly and
undignified violence of language," fretted the *New York Times*, "it sur-
passes all his previous achievements, and by its tone and temper it

engenders a natural apprehension as to the extremes to which this ill-balanced man may permit his perfervid zeal to carry him."

Elsewhere in its pages, as if to drill home the point, the *Times* quoted Chancellor Day of Syracuse University, who characterized the president's message as "the ravings of a disordered mind." The *New York Sun* gravely concurred. "It is an even more disturbing reflection that the hand which penned this message is the same hand which directs the American Navy, now on its mission toward unknown possibilities. God send our ships and all of us good luck!"

The following afternoon, while Americans were still digesting the ravings of their president's mind, the jury in the trial of Harry Thaw returned a verdict of not guilty by reason of insanity. The moment the verdict was read, a young spectator named Theodore Roosevelt Pell—a distant relative of the president who had apparently inherited the same tendency for perfervid zeal—shouted out his approval. The judge fined Mr. Pell $25 for contempt of court.

Meanwhile, Harry Thaw's own reaction suggested the jury had made the right decision. At first, he was jubilant, so dizzy with relief, in fact, he seemed not to hear the judge's order that he be immediately remanded to Matteawan Asylum for the Insane near Fishkill, New York. Once he did comprehend the order, his glee dissolved into a tantrum. "I will not go to Matteawan," a reporter overheard him say to his attorneys. "I will not go there."

"Where did you think you would go," wondered one of his attorneys cruelly. "To Rector's or Martin's?" In his earlier life, Thaw had frequented both of these fashionable restaurants.

Late that afternoon, accompanied by police escorts and pursued by the press corps and curious spectators, Thaw and his wife were driven to Grand Central Terminal in Evelyn's electric automobile. With a word, they parted, Evelyn to return home to Park Avenue, Harry to board a private rail car on the 4:39 to Fishkill. The *Times* expressed satisfaction that this "dangerous and degraded being" would never be free again.

A Journey for Madmen

As the sage editors of the *New York Times* gazed out over the world from the high windows of their twenty-five-story skyscraper, passing judgment on the deranged men who made news beyond their Times Square doorstep, they might have paused to wonder whether their steel-framed edifice was not, in fact, a kind of glass house; and whether they were in any position to be casting stones.

In late November of 1907, the *Times* had agreed to cosponsor, with the French newspaper *Le Matin*, a twenty-thousand-mile automobile race from New York City to Paris, via the continental United States, Alaska, Siberia, and Europe. The *Times* assumed responsibility for arrangements as far as the Cape Prince of Wales on the Bering Strait; *Le Matin* would oversee the way from Siberia to the center of Paris. The race had been hastily planned and was now scheduled to commence from Times Square on February 12, Lincoln's birthday. As of early February, five foreign auto crews—three from France, one from Italy, another from Germany—had officially entered the race, but not one American automobile manufacturer had committed to sponsoring a contestant. All had apparently concluded that a twenty-thousand-mile race around the world was a fool's errand.

Fool's errands were nothing new to newspapers at the start of the twentieth century. American journalism could boast a long history of funding fabulous, if dubious, stunts, then reporting on them as world-shaking exclusives. James Gordon Bennett, publisher of the *New York Herald*, had profitably sent Stanley to find Livingston in the jungles of Africa. Joseph Pulitzer had sent Nellie Bly to travel around the world in less than eighty days (thereby beating the fictional record set by Phileas Fogg of Jules Verne's *Around the World in Eighty Days*) for his *New York World*. Many newspapers, too, had sponsored automobile races in the early twentieth century.

But this was no ordinary automobile race; and for sheer outlandishness, it rivaled anything the *Times*'s competitors had attempted. The *Times* declared the automobile race from New York to Paris "the longest and most perilous trip ever undertaken by man." One of the contestants described it, more accurately, as "a journey for madmen."

Madness is the last thing most readers would have associated with the *New York Times* in 1908; another paper, maybe, but not the *Times*. Since purchasing the then-floundering *Times* in 1896, publisher Adolph Ochs had endeavored to position his newspaper as the restrained, conservative alternative to the baser offerings of Hearst's *Journal* and Pulitzer's *World*, or even Bennett's *Herald,* among other of the city's sixteen dailies. His was to be the journal that "would not soil the breakfast cloth," as he once promised. Rather, it would be "a clean, truthful, carefully edited newspaper ... that recognizes its obligation to give its readers all the news, but values its own good name and their respect too highly to put before them the untrue or the unclean or to affront their intelligence or their good taste with freaks of typographical display or reckless sensationalism."

Ochs found a readership that liked what he offered. By 1908, circulation had risen to a very respectable 173,000, and the *Times* was in a position to congratulate itself for its foresight and righteousness. "In the City of New York, they argued, a newspaper could secure a large circulation only by giving itself over to sensationalism and pictures, to startling headlines and yellowness," a 1908 editorial recalled. "The management of THE TIMES was not of that opinion."

Actually, the opinion of the *Times* management was a little more complicated than statements like this implied. Ochs's genius as a publisher was to combine dignified sentiments with sharp instincts for what sold papers, even if the latter occasionally trumped the former. Seldom did an issue of the *Times* run off the Hoe presses in the subbasement of the Times building without containing one or two frivolous but eye-catching human-interest stories, exactly the "freaks of typographical display" the paper swore to disdain. Thus, that winter, *Times* readers learned of a woman in New York who dug up her husband's corpse in order to have her photograph taken with it; of a man who laughed himself to death and of another who went blind after a dream; and, most poignantly, of the forlorn Mr. Cashetto of Youngstown, Ohio, a prosperous Italian who went insane one February afternoon after walking into a nickelodeon and seeing a beautiful young woman appear on-screen. "Cashetto uttered one scream," reported the *Times.* "The woman was his long lost wife."

Since moving into his new headquarters in 1904, then the second-tallest skyscraper in the city—Ochs, counting basement floors, insisted it was the tallest—the publisher had made sure his new building stood at the center of the ever-expanding city, attracting crowds to the newly christened Times Square to witness marvelous displays of fireworks and electric lights. His latest ploy of dropping an enormous electric ball from the building's flagpole on New Year's Eve had drawn tens of thousands.

And now he would draw them again, this time to witness the start of something even more spectacular: *La coupe de monde,* as the French were calling it. The race around the world.

With less than two weeks to go, and the European contestants already bound for New York, plans for the race were still coalescing. Its general route, at least, was set. From Times Square, the contestants were to drive northwest across the continental U.S., by way of Buffalo and Chicago, then west through Wyoming and Utah, over the Sierra Nevada and into California. This leg of the journey would in itself qualify as a noteworthy undertaking in 1908. Only nine times had an automobile crossed the continental U.S., and just one of these crossings had occurred during winter.

As originally conceived, the plan called for the contestants to turn north into Canada and proceed to Alaska through British Columbia, but by early February the Canadian leg had been dropped. The contestants would travel instead by ship from San Francisco to Valdez, Alaska. The use of ships had been expressly disavowed in early plans for the race—"Sans prendre le paquebot," *Le Matin* had stipulated—but no one now objected to this abridgement.

From Valdez, the racers were to drive north along dogsled trails to Fairbanks, then take to the ice of the Yukon River and drive west to the Bering Strait. Assuming all went well, they would finally arrive at Cape Prince of Wales in early spring. Skirting the Arctic Circle, they would drive across the ice of the frozen strait and enter into "the trackless, frozen wastes of Siberia, a land so forbidding and repellant to comfort-loving man," the *Times* described it. The racers would proceed west through Siberia, west and west again, circumnavigating the North Pole through Russia, Germany, and France. They would eventually arrive, according to the plan, in the center of Paris sometime in the summer of 1908.

Putting aside the fact that few automobiles in 1908 were equipped to handle the snows of Central Park, much less of Alaska, or that few automobiles could mount a really steep hill, much less the Rockies or Sierra Nevadas, the route presented certain obstacles. One experienced traveler to Alaska, C. H. Rogers, prophesied in a letter to the *Times* that these obstacles would become more or less insurmountable as its route veered north. "In the first place, let us take the road proposition," wrote

Mr. Rogers. "This one item will prove my assertion that no auto will reach the Bering Strait, or within 1,000 miles of it." Rogers pointed out that dog team trails, the only open paths through the snows of Alaska, were two feet wide and banked by high soft snow on either side. "I believe in progress," Mr. Rogers assured the *Times*. "I believe airships will soon be as common and useful, if not more so, than the auto, but never, never, will an auto reach Bering Strait with its own power over a dog trail."

Of course, getting to the Bering Strait was only part of the problem. Once the racers arrived, they were expected to drive across it. The Bering Strait was a fifty-mile-wide neck of salt water connecting the northern Pacific and the Arctic Ocean that made the Strait of Magellan look positively snug by comparison. Despite average winter temperatures of around zero degrees Fahrenheit, the strait had not frozen over completely in twenty years.

In reality, the racers had about as much chance of driving through Alaska and across the Bering Strait as they did of flying to the moon, but the *Times*, cheerfully, if not willfully, oblivious of this reality, forged ahead with the plan. "If the machines could land in Alaska before April 1 the chances of their getting through that country by the first of May are excellent, avoiding the thaw all the way," the *Times* postulated. "It is believed that the cars can cross Bering Strait on the ice as late as May 5 to 10, and that they may be able to reach East Cape for the Trans-Siberian journey before the tundra thaws in the interior."

That the staid and serious *New York Times* would embrace this quixotic venture was puzzling on the face of it; the *Times* had apparently let credulity and excitement overcome its usual reserve. In the context of 1908, though, the paper's sponsorship of the race around the world did make a kind of sense. For one thing, the conviction of the *Times* that the route was feasible reflected an ignorance of geography—not to mention meteorology—shared by most Americans and Europeans at the time. Large areas of the earth's surface remained terra incognita to Westerners in 1908. Not only had both the North Pole and South Pole eluded man, but vast swatches of eastern Europe, including much of the Balkans, were unexplored by outsiders. So, too, was northern Siberia, the central Asian desert, the East Indies, central Australia, three-fourths of Canada, and huge sections of the Amazonian rain forest. Even parts of the eastern United States remained practically untrod by nonnatives. "The southern half of Florida is still the haunt of the Indians, who live there much as in De Soto's time," stated *Harper's Weekly* in 1908. "Palm Beach and the most fashionable resorts of the east coast are hardly more than oases along the edge of the pathless wilderness of Mangrove swamps."

If believing an automobile might drive across Alaska and the Bering Strait was a symptom of ignorance, it was also a testament to the faith Americans and Europeans put in man and his new machines in the early twentieth century. Adolph Ochs and his managing editor, Carr Van Anda, subscribed to this faith wholly and sincerely. Granted, the race was an opportunity to boost circulation and tap into a burgeoning ad market for automobiles—the publisher was too canny a businessman to miss this—but it was also a chance for the *Times* to partake in a glorious story of human and technological achievement.

The race neatly combined two popular fascinations of the day. The first of these was, of course, the automobile. Many Americans, even those who claimed to despise the machines—and those who could not afford them—were beguiled by their potential. Just how fast and far could they go? Track racing, beach racing, long-distance endurance runs, hill climbs—all had become instantly popular sports, as Americans attempted to find out. The race to Paris would encompass all of the above challenges, plus the added challenge of traveling over hummocky ice through subzero temperatures. "[T]he prospect of its success is staggering," the *Times* declared, "a conception worthy of a [Jules] Verne which, if achieved, will upset all ideas of the limitations of motor car travel and will show to the world that here at last is a power greater than any of the wonderful means of transportation which human ingenuity has yet given us."

But this would be much more than a test of machine. It would require humans to endure exquisite cold in desolate parts of the world. This added dimension tapped into another subject of intense interest and speculation at the start of the twentieth century: the North Pole and the attempt by Americans to claim it. The contestants of the race would not be going *to* the North Pole, of course, just *around* it. But this was close enough to inspire comparisons in the press. "The stupendous undertaking is almost on a par with the dash for the pole," the *Baltimore American* stated of the race. "It shows how deeply seated the spirit of adventure and of emulation is in the human race."

Signs of that spirit were everywhere at the start of the twentieth century. It was the last great age of exploration—the "Heroic Age," it is sometimes called—and no earthly goal was nobler than the attempt to conquer unknown and primitive parts of the world, or die trying. No fewer than nine major exploratory expeditions from Europe and America were on the verge of launching, or were already launched, in early 1908. In addition to expeditions to the Brazilian Amazon, the Himalayan Mountains, and the central desert of Australia, four expeditions were bound for the Antarctic, most notably the British Antarctic Expedition led by Ernest Shackleton, whose *Nimrod* had departed Christchurch, New Zealand, on New Year's Day, 1908.

But it was the northern latitudes that exerted the strongest pull on Americans' imaginations. For well over a century, American explorers had been trying to reach the North Pole, always failing, often dying, and yet persistently returning. Exploration of the far north combined the shivery allure of vast icescapes with the satisfying prospect that civilized man would soon bring this region of the earth under his dominion. "Beyond the geographical and scientific value of the discovery of the North Pole, and the solving of questions of popular curiosity," the American explorer Anthony Fiala wrote in 1907, "another reason exists to explain the ceaseless effort to reach that mystic point: the Spirit of the Age will never be satisfied until the command given to Adam in the beginning—the command to subdue the earth—has been obeyed, and the ends of the earth revealed their secrets."

At the start of February 1908, Dr. Frederick Cook, of Brooklyn, New York, was wintering in a small Eskimo settlement in northern Greenland, waiting for the sun to return to the far north so he could begin a thousand-mile dash to the North Pole in hopes of claiming it, at last, for America. In New York City, meanwhile, Robert Peary was making the rounds of the city's newspapers, including the *New York Times,* attempting to secure funds and publicity for his own return to the far north. Peary had already gone to the Arctic seven times and tried for the Pole twice. He had lost eight of his toes to frostbite and sacrificed the better part of his adult life to his pursuit. In July, he would try again.

Given Americans' general enthusiasm for northern exploits, the absence of an American car in the race around the pole was conspicuous. Moon Motor Car Company of St. Louis had announced its intention to enter one of its Hol-Tan cars back in late November but then dropped out. Two other companies, Maxwell and White, had likewise flirted

with the race then withdrawn. The E. R. Thomas Motor Company of
Buffalo was still a possible contender but had so far made no commit-
ment. None of the large automobile manufacturers, such as Packard,
Pierce-Arrow, or Oldsmobile, had apparently even entertained thoughts
of entering the race. That three of the five automobiles included were of
French origin seemed not only to confirm the conventional wisdom
that France dominated the car industry, but to suggest that American car
manufacturers—if not Americans in general—were a timid lot, beaten
before the race began. This was clearly not the story the *New York Times*
had intended to report when it agreed to cosponsor the race.

By the start of February, the European race crews were somewhere out in
the middle of the Atlantic Ocean, crossing to New York in seas so heavy
many of the contestants took to their cabins in distress. Even seasick, they
might have been interested to know that Ernest Shackleton's *Nimrod* had
just steamed through McMurdo Sound and was currently wedged among
cakes of ice at the edge of the Great Ice Barrier of Antarctica. They might
have been further interested to know that, in addition to ten Manchurian
ponies and eight Siberian huskies, the British explorer had taken with him
one Arrol-Johnston motorcar. Shackleton had brought along the motorcar partly as a practical measure to help with hauling and transportation,
but also as a chance to test the operation of automobiles in extreme
cold—the sort of cold the New York–to–Paris racers could expect to
encounter on the Bering Strait in late winter.

On February 1, a balmy late summer day in Antarctica with temperatures hovering around fifteen degrees Fahrenheit, Shackleton and his
men rolled the automobile off the *Nimrod* and onto the ice pack. After a
bit of tinkering, the automobile was ready for its polar debut. The motor
turned over, the car chugged along a hundred yards or so, and then, as its
tires sank into the soft ice, came to a halt and refused to budge further.
Shackleton ordered the ignominious machine loaded back onto the
boat.

The Thomas Flyer

A frigid blast of air hit New York at the start of February. Temperatures
in the city dropped to one degree Fahrenheit. Upstate, in the Adiron-
dack Mountains, thermometers recorded lows of minus fifty. By Febru-

ary 4, an ice bridge had formed across the Hudson River at Nyack, thirty miles north of the city. Great floats drifted into New York Harbor and bunched up against the shores of the Verrazano Narrows, giving the city an appropriately arctic cast as the European contestants of the New York–to–Paris race entered New York Harbor.

The contestants began to arrive on Friday, February 7. First came the Germans aboard the *Kaiserin Augusta Victoria*, accompanied by their bulky Protos motorcar. The three-man team was led by Hans Koeppen, a thirty-one-year-old lieutenant in the Fifteenth Prussian Infantry. He was a man of impeccable pedigree, robust and handsome, all qualities that nearly made up for the fact that he had no idea how to drive an automobile. Most of the actual driving would be done by his teammate, Hans Knape, a twenty-eight-year-old Berliner best known in his country for the skills and daring he'd displayed in the lethal new sport of motorboat racing.

The following day, the French and Italians arrived with their automobiles, having sailed together from Le Havre aboard the *Lorraine*. As the racers gathered on the quarterdeck to have their portrait taken by a *Times* photographer, they presented a colorful if motley crew of adventurers, poets, and blowhards. The personality trait they all shared was a willingness to endure untold misery for the sake of dubious glory. Three of the Frenchmen had participated in the nine-thousand-mile Paris-to-Peking race the previous year; two of these men, August Pons and Charles Godard, had nearly died in the Gobi Desert during that race. They, at least, had some idea of what to expect in terms of suffering; then again, the New York–to–Paris race promised to be twice as long and many times more difficult.

On the same day the Europeans were acclimating themselves to the skyscrapers of New York City, the Great White Fleet was threading its way through the snowcapped peaks that edged the Strait of Magellan. The fleet had anchored for several days mid-strait, in the austere Argentine port of Punta Arenas. At 11 P.M. on the night of February 7, the battleships set sail again, heading southwest. Just as dawn broke, the fleet rounded the lowest point of the South American continent, then turned northward to negotiate the narrowest channel of the strait. The sun rose to reveal precipitous black rocks jutting out of water that ran as deep as three thousand feet. Waterfalls dashed down the cliffs and fierce shocks of wind, called "willywaws," swept off the rocks and sent shudders through the ships and men. A line of dark swells stretched across the channel, jostling the ships as they left the waters of the Atlantic Ocean for the waters of the Pacific. By eight in the evening, the Great White Fleet had passed Cape Pilar and was steaming north, no worse for its week in the notorious Strait.

Good news for the fleet; not so good for the shipboard journalists, who by now were probably beginning to suspect they'd been sent far from home to cover the greatest anticlimax in naval history. "But you want to hear all about it?" Franklin Mathews, correspondent from the *New York Sun*, asked readers almost plaintively. "Is there an impatient call for details of this much-heralded trip of dread, a breathless demand to know how many close calls and narrow escapes there were from hitting sunken rocks, gliding against precipices, scraping the paint from ships' sides, dodging willywaws?

"Well, if you guessed any or all those things you must guess again. None of 'em happened."

Not that anyone that weekend was too concerned about the fate of the fleet. The attention of New Yorkers was fully absorbed by race preparations. The arrival of the foreign drivers, clad in arctic garb of white fur and black rubber, had conferred upon their venture a sudden exhilarating reality.

On Sunday, February 9, a frigid but cloudless day, the city woke in a festive mood. Five thousand skaters rushed to Central Park after church services to glissade over the thick ice on the lake. Sleighing parties coursed the lanes and trails, breaking into illicit impromptu races, or "brushes," when police were not around to intervene. At the southern tip of the island, the harbor remained choked under the heaviest ice pack in years, bringing ship traffic to a halt. Near Sandy Hook, at the entrance of New York's harbor, hundreds of passengers climbed down from ice-locked ships and walked half a mile across solid ice to the shore of New Jersey.

That afternoon, the foreign race contestants drove out along the Hudson River in a convoy to survey Westchester County. Most of the foreigners had never been in New York before, and as their cars chugged up Broadway, they gazed with wonder at the towering apartment buildings, and across the ice-choked river at the snow-streaked Palisades rising from the opposite shore.

And then came news to add to the high spirits: the Americans were in the race. The E. R. Thomas Motor Company of Buffalo, New York, after weeks of hemming and hawing, had committed to entering one of its Thomas "Speedway Flyers." The four-thousand-dollar runabout was to be loaded onto the flatcar of an express train in Buffalo that same evening and would arrive Monday morning in New York. Montague Roberts, a celebrated and dashing twenty-six-year-old professional race-car driver, had been engaged to drive the automobile, an added attraction to male and female race fans alike. The *Times* predicted that Roberts's progress "will be watched with the keenest interest by every American with a drop of sporting blood in his veins, whether he ever enjoyed the exhilarating experience of motoring in the open air or not."

The morning of Wednesday, February 12, Lincoln's birthday, brought warmer temperatures and mild breezes. The day was a public holiday, and the streets remained quiet through the early hours. Under long shadows, Times Square fluttered with tricolored bunting and flags that streamed from the empty grandstand and surrounding buildings. By ten o'clock, with the temperature just below freezing, the streets started to fill. By eleven, a crowd of fifty thousand was gathered within Times Square, and tens of thousands more were assembling along Broadway to the north.

At the center of the crowd's attention was the pack of six automobiles parked in front of the Hotel Knickerbocker. The entries ranged from the German crew's boxy Protos, a six-thousand-pound, forty-horsepower wagonlike vehicle covered in canvas, to the tiny French Sizaire-Naudin, a pipsqueak weighing in at a mere thirty-three hundred pounds and containing a single-cylinder twelve-horsepower engine. The Thomas Flyer packed the most punch per pound. It weighed only a thousand pounds more than the Sizaire-Naudin but carried a seventy-horsepower, four-cylinder engine.

More evocative than the automobiles were the burdens they carried. Each was loaded with a small arsenal of rifles and revolvers to defend the men against wolves and marauders, or provide food, as necessary. Other equipment included ropes, picks, shovels, axes, hatchets, crowbars, blowtorches, wrenches, wheel chains, wooden planks, lanterns, searchlights, compasses, sextants, and assorted thermometers, barometers, and hydrometers. Most of the automobiles carried extra tanks for gasoline. One, the French De Dion, came equipped with a mast and sail.

Just before eleven, the automobiles moved into their assigned positions in the middle of Times Square, side by side, facing the Times building to the south. The Thomas Flyer slipped between the French Motobloc and the Italian Zust. Montague Roberts was at the wheel, sitting at the right side of the car, where steering wheels were still customarily located. At his left sat a heavyset unremarkable-looking man named George Schuster who, until the previous morning when he received a call from Buffalo inviting him to join the race, had been a mechanic and road tester for Thomas Motor Company. Schuster had arrived in New York by train that very morning from Rhode Island. Stuffed into the rear bucket seat was T. Walter Williams, a forty-three-year-old reporter for the *New York Times* whose colorful past as a gold prospector, merchant marine, and Nile boatman made him an ideal candidate for this unusual

The Thomas Flyer in Times Square on the morning of February 12. Montague Roberts sits behind the wheel, on the right side of the car. Not until later in the year would the wheel move to the left in American cars.

assignment. Williams's daily dispatches, sent by telegraph from posts along the route, would be published in the *Times* and disseminated by wire service to newspapers throughout the country.

The official starting time was 11 A.M., but the hour passed and still the motorcars waited. Mayor George McClellan, the official starter of the race, had yet to arrive and take his place on the grandstand. This being Lincoln's birthday, a few people in the crowd might have recalled the mayor's father and namesake, General George B. McClellan, who'd led the Army of the Potomac at the start of the Civil War. It was McClellan senior whom President Lincoln, having grown weary of the general's reluctance to advance on the Confederate army, once accused of suffering a case of "the slows." Apparently, the apple did not fall far from the tree.

And so they all waited for McClellan junior to arrive: the racers bundled in their arctic gear among their arsenals; the fifty thousand spectators shuffling and stamping and craning their necks to glimpse the men. The automobiles idled, their mufflerless engines drowning out the brass band. "American flags fluttered in the gasolene-scented breeze," noted the *New York Tribune*. A quarter of an hour passed. Finally, at 11:16, a man named Colgate Hoyt, president of the Automobile Association, stood up in the grandstand and reached for the starting pistol. He made a motion with his hand, then fired.

The automobiles began to move, one following the other, first lurching downtown on Broadway for a block before rounding the Times building and turning onto Seventh Avenue, which led them back into Broadway, uptown now, northward. Mounted police cleared a narrow lane through the crowd, giving the automobiles just room enough to pass. Behind the six contestants, several hundred additional automobiles, by prearrangement, pulled in behind the racers to follow them to the outskirts of the city.

For half an hour, the procession was more of a stately parade than a race. The blocks accumulated into miles, but the crowd did not slacken. In Morningside Heights, almost four miles from Times Square, hundreds of students from Columbia University ran from classes to watch the cars pass and cheer them on. The racers dipped down to 125th Street, then rose again toward Harlem Heights. Moments later, they were gone over the crest of the hill, bound for Paris by way of the far side of the earth.

Shortly after 11 A.M. on February 12, the race contestants head up Broadway from Times Square, off to Paris.

CHAPTER FOUR

Middle Earth and the Night Riders

*They are anarchy materialized; and if they are to flourish, civ-
ilization in Kentucky must perish.*
—*The Washington Post,* February 18, 1908

The same sun that shone on New York City so brightly that morning in February never appeared above the small Eskimo settlement of Annoatok, three thousand miles to the north, on the western
coast of Greenland. The sun malingered beneath the horizon, casting a
few weak rays skyward to produce a feeble twilight before receding again
into the polar night. Dr. Frederick Cook stood in the icy air and studied
the sky for signs. He watched the sun's ambient light extend a little further each day, bleeding in from the south. He observed the color of the
snow as it shifted subtly from deep bruised purples toward dark violets.
One morning, soon, the sun would rise, and then he would begin his
thousand-mile trek to the North Pole.

Frederick Cook had arrived at Annoatok five months earlier, in September of 1907, with a boatload of supplies and dogs. In the relative
comfort and warmth of his stove-heated cabin, he'd spent the dark winter preparing for his journey. The first important step had been to make
the wooden sledges that would carry the expedition's supplies. With the
assistance of the Annoatok Eskimos and a young German named
Rudolph Franke, Cook had constructed eleven of these sledges, including one that could be converted into a small boat. He'd also arranged for
Eskimo hunting parties to advance northwest of Annoatok and lay in

caches of musk ox and deer along the route to the Arctic Ocean, ensuring fresh meat for the first leg of his journey. This fresh meat would be
supplemented, and later replaced entirely, by a large supply of the ox-beef
and seal-meat pemmican that Cook and the Eskimos had been drying
through the winter.

During moments of leisure that winter, Cook socialized with the
natives. He had taken time to learn their language, which not only facilitated his work with them but also gave him entry into their customs and
beliefs. A generation earlier, Victorians had considered Eskimos as raw
and filthy primitives. (Charles Dickens dismissed them as "a gross handful of uncivilized people, with a domesticity of blood and blubber.")
Cook, too, patronized the Eskimos as "faithful savages," but he genuinely
admired their intelligence and stoicism, and passed no judgment on
their ways of life. Infanticide and wife swapping were two Eskimo customs not likely to go over well back home in Bushwick, Brooklyn, where
Cook's wife and children awaited his return.

As the moment of departure drew near, Cook tested his equipment
and reviewed his plan of approach. He knew from experience how critical his preparations would be to his survival in the polar wilderness,
where the margins of error were slender. He had been to the far north
several times, twice on expeditions led by Robert Peary and once to
climb Mount McKinley in Alaska. He'd also distinguished himself as the
first man to visit both the Arctic *and* the Antarctic when he joined an
expedition to the South Pole in 1897. He knew of the countless would-be
heroes who had gone to the far north or far south since the mid-
nineteenth century only to be turned back, or killed, by their failure to
comprehend the exigencies of polar exploration.

Such endeavors, by their very nature, attracted men who were remarkably courageous and bold, but also prone to delusion. Given the slim odds
for success, a certain amount of delusion was probably necessary in a polar
explorer. They all required the capacity to believe unbelievable things. An
old friend of Cook's, the Norwegian Roald Amundsen—who was famed
for discovering the Northwest Passage in 1906 and who would become the
first man to reach the South Pole in 1911—had sounded positively fanci-
ful when he announced, on a visit to New York in October of 1907, his
own peculiar designs on the North Pole. He planned to have himself
pulled by trained polar bears. "It is quite true," Amundsen assured the *New
York Times*. "Each bear will draw a sled. They will be fed on seal meat."

Cook was prone to delusion himself. And like all explorers, he was vain enough to cast himself in the role of conquering hero while masochistic enough to invite the suffering the task entailed—to wear a crown of thorns in exchange for a crown of laurels. But Cook was also methodical and pragmatic. A physician by occupation, he was a man of science, trained to consider a problem rigorously and objectively. This is how he approached the challenge of reaching the North Pole.

The "problem of the Pole," as explorers sometimes referred to it, came down to food. Walking a dozen miles a day over snow and ice, in temperatures between minus thirty and minus eighty degrees Fahrenheit, men and dogs required large quantities of food to fuel their bodies. But the more food they carried, the heavier their loads; and the heavier their loads, the greater their expenditures of both energy and time, which in turn necessitated *more* food. When newspaper writers applied the term "race" to polar expeditions, they made it sound as if getting to the North Pole were merely a matter of sporting competition. Competition certainly existed—Cook was knowingly attempting to reach the Pole before his old commander, Robert Peary, could claim the trophy—but the real race was against time. Food was the sand in the hourglass. When it ran out, time was up.

The nineteenth-century approach to polar exploration, best exemplified by Great Britain's doomed Franklin expedition of 1845, was to penetrate the Arctic with massive amounts of food, supplies, and manpower, invading the far north like a Roman army. Cook favored the more modern approach pioneered by several Norwegian explorers. His plan was to move quickly with a small force, starting with ten men, then paring down to three men—himself and two Eskimos—for the last dash across the ice. He'd made his sledges out of hickory, a light but strong wood he'd brought with him from home. Light sledges would skim fast over the ice, lessening the burden on the dogs, and therefore reducing the amount of food they would require.

Cook intended to pursue a route to the Pole no explorer had tried before, hewing to land as long as possible to benefit from the possibility of game. The more fresh meat they could hunt early in the journey, the better the men's chances of survival later. The sledge dogs would have fresh meat, too, much of it provided by themselves. As the journey progressed, the weaker dogs would be slaughtered to feed the stronger. The tactic was not likely to win Cook adoration among members of the

Association to Prevent Cruelty to Animals back in New York, but it was a brilliantly efficient method of transporting meat on its own four legs. Polar exploration was literally a dog-eat-dog pursuit.

On the morning of February 19, Cook saw the sign he'd been looking for. The sun glimmered in the south, inching above the horizon and casting "a belt of orange" over the southern sky. The temperature rose to a relatively toasty minus thirty-six Fahrenheit; the air, if not quite scented with spring, carried at least a hint of sunnier days to come. Cook ordered the sledges packed. In addition to the supplies they would need along the way, such as lamp fuel and navigational instruments, the men loaded nearly a thousand pounds of pemmican. For warmth, they wore blue-fox coats, birdskin shirts, bearskin pants, and sealskin boots. The dogs, 103 of them, well-fed and yapping excitedly, were harnessed to the sledges. Fortunately for all concerned, they had no idea what awaited them.

Cook was likewise ignorant of the future, of course, and probably better for it. Once he stepped onto the ice of the Arctic Ocean, he would be treading into places no human, white or Eskimo, had ever been. It was impossible to predict with any certainty what he and his men would find when—if—they got to the Pole, a great white canvas onto which people could project whatever fantasy they wanted. The ancient Greeks had imagined the Pole to be a temperate paradise with an open sea at its center, a theory bolstered in the nineteenth century by scientific observation. The Eskimos called the Pole the "Big Nail," because they believed an enormous iron spike stuck out of it like an obelisk. Children thought it was the home of Santa Claus.

The most intriguing theory of the North Pole was forwarded by a New York City organization called the Hollow Earth Exploring Club. According to the president of this club, William Reed, the North and South Poles were both gaping holes that gave into the interior of the earth and met in its middle, forming a tunnel straight through from north to south. This hole, according to a speech Mr. Reed gave at the club's first public meeting in the early spring of 1908, was the original source of such natural anomalies as icebergs and penguins. It also accounted for the mysterious disappearance of explorers who had gone missing in the far north. The old explorers were not dead, insisted Mr. Reed. They were living happily in the middle of the earth.

Frederick Cook never did reach Middle Earth. But where he ended up would be, in its way, every bit as strange.

The Black Patch

As Frederick Cook began his journey to the north, the contestants of the New York–to–Paris race struggled through the early stages of their own northern trek. Almost immediately after leaving New York City and entering the suburbs of Westchester County, they had been rudely introduced to the sorry state of American roads. Groomed avenues gave way to narrow lanes more fit for mules and horses than automobiles. Unplowed roads vanished into snowdrifts, forcing the race crews to stop and shovel their way out before proceeding. Occasionally, the racers abandoned the road altogether in favor of fields and orchards. They did this either intentionally because the road was impassable or accidentally because they could not find the road under the snow and simply took the path of least resistance.

Montague Roberts, the handsome young American driver, grabbed the lead for the Thomas car early and held on to it as the others followed in hot pursuit—at about twelve miles per hour. Holding the lead was a mixed blessing, since it fell upon the leaders to do most of the shoveling. Somewhere west of Schenectady, a helpful local suggested the racers try the towpath of the Erie Canal. The surface of the ten-foot-wide path was scoured by the wind, making for an icy but clear passage. For the first time since leaving New York City, the racers were able to attain real speed, albeit at considerable risk. A single false move could send them tumbling down into the frozen canal and crashing through the ice. By February 19—the same day Frederick Cook departed from Annoatok— the Thomas car was clear of the states of New York, Pennsylvania, and Ohio and was heading into Indiana and one of the greatest blizzards in that state's history.

The route the racers traversed in that first week took them through great, modern, highly industrialized cities of the north, like Buffalo, New York (home of the Thomas Motor Company), and Cleveland, Ohio. The sorry condition of the nation's roads brought home a different reality. America was in some ways a land as primitive as any the racers would travel through on their way to Paris. Whatever popularity automobiles enjoyed among the affluent in cities and suburbs, they were still rare curiosities in the countryside, outnumbered by horses a hundred to one. And despite advances in farm equipment and new services like rural free delivery, telegraph and telephone, and Sears catalogues that offered everything from the latest fashions to prefabricated houses—despite all these modern amenities, the rural interior was still largely premodern. More than 90 percent of Americans lived without electricity in their homes. Personal hygiene and public education were irregular, medical care rudimentary, indoor plumbing a pipe dream.

It was in its tendency for violence, though, that America revealed its primitivism most clearly. Not just in jarring splashes of murder or assault, but in festering campaigns that threatened the very idea of America as a place of law and order. The violence was facilitated by the prevalence of guns in the country. The racers, well armed themselves, became quickly acquainted with America's love of firearms. They were often greeted by blasts of salutary gunfire as they passed by a farmhouse or entered a rural town. It was not always clear whether the gunfire was meant to say hello or, more convincingly, good-bye.

Nowhere was the American taste for violence more dramatically displayed in the winter of 1908 than in an area of the country a few hundred miles southwest of where the racers were traveling. Among the gently rolling hills of western Kentucky and Tennessee, in a cluster of rural counties known collectively as the Black Patch—so-called for the dark-leafed tobacco that grew in its soil—armies of men on horseback galloped through the countryside under cover of winter night, setting fire to tobacco warehouses and destroying tobacco fields, whipping and shooting those who crossed them. The men called themselves the Silent Brigade. Most Americans knew them better as Night Riders.

The Night Riders were a well-armed, well-organized insurgency of tobacco farmers and other local men formed to defeat the so-called tobacco trust. Like just about every other industry in America in 1908, tobacco was controlled by trusts, most completely by James B. Duke's American Tobacco Company. In the Black Patch, as it happened, American Tobacco played a relatively small role, but the situation was much the same as in other tobacco-growing communities. Most of the crop in the region was purchased by a consortium of foreign buyers, mainly Italians, who savored its thick, fire-cured leaf. The agents of these foreign buyers had colluded, trustlike, to fix the price for tobacco at about four cents a pound, which was a penny or two lower than the cost of growing it. For the small tenant farmers who lived in these parts of lower Kentucky and upper Tennessee, and who had depended on tobacco as their money crop since their ancestors migrated over the Allegheny Mountains in the late eighteenth century, the low price imposed a hardship they came increasingly to resent. Starting in 1906, they turned to force to resist it. And by the end of 1907 and start of 1908, they were conducting a series of breathtaking raids to destroy the grip of the trust and raise the price of tobacco.

The raids followed a similar choreography. After advance scouts sneaked into a sleeping town to perform midnight reconnaissance, squads of riders swept in and seized the district with martial efficiency, cutting phone lines, blocking streets and train tracks, forcibly subduing police or citizens who might cause trouble, then torching or dynamiting the property of their foes. Tobacco warehouses ignited in magnificent pyrotechnical displays, as tens of thousands of pounds of oily tobacco leaf went up in flame and smoke. Afterward, Night Riders were seen galloping home through the dark countryside, singing Christian hymns as they rode.

The most recent raid had occurred in the early-morning hours of February 16, when 250 riders converged on the town of Eddyville, Kentucky, along the banks of the Cumberland River. The Night Riders attacked the town at about 1 A.M. After commandeering the telephone office and forcibly procuring a dozen horse whips from the local hardware merchant, the riders rounded up ten men—four white, six black—and took them to the banks of the river. Under a nearly full moon, they whipped the men until the skin shredded off their backs.

By such tactics, Night Riders had effectively imposed a reign of terror over the Black Patch that few local judges, lawyers, or constables dared to defy. Tobacco growers who sold their leaf to the trust found their fields raked or dynamited, their homes shot up, their backs bloodied. The man who served as the governor of Kentucky until December of 1907, J.C.W. Beckham, had apparently calculated that the effort required to intervene was worth neither the political nor personal risk. The new governor as of January 1908, Augustus Willson, conceded that Night Riders, ten thousand strong, effectively controlled a third of the state. Night Riders were also making incursions into the tobacco states of Virginia and North Carolina. The governor of Ohio, fearing his own state might be attacked, had posted National Guard troops along the Ohio River to prevent approaches from the south. Riders managed to cross the river anyway and wreak havoc before slipping back into Kentucky.

Tales of the Night Rider "outrages" sent shock waves through the country and across the Atlantic to Europe, particularly after the Italian government began demanding reparations for destroyed property. Journalists from the *Saturday Evening Post* and *Harper's Weekly*, among other national publications, traveled to Kentucky and Tennessee to investigate the terror, while newspapers throughout the country ran frequent dispatches. Altogether, the stories painted a harrowing picture of a piece of America overtaken, as the *New York Herald* put it, by an "unexpected, mysterious, mythical force."

"They are," the *Washington Post* declared of the Night Riders, "anarchy materialized."

Anarchy materialized in a great many forms that year. While Night Riders presented an especially haunting and picturesque specter of marauding horsemen setting towns ablaze by moonlight, most Americans could find comparable, if more prosaic, incidents of terror right outside their own front doors. In New York City, loosely organized gangs of Italian extortionists, known collectively as the Black Hand, routinely dynamited the homes and businesses of fellow Italian immigrants. Bombs exploded almost nightly in Little Italy, sending scores of panicked Italians running from burning tenements in their nightclothes screaming, "Mano Nera, Mano Nera." In the south, lynching was epidemic. American black men were routinely beaten and hung from trees that year.

Alongside the criminal and racial outbursts, much of the violence in America was political. The long war between capital and labor that had begun in the latter half of the nineteenth century was escalating to a new, more bellicose phase. Trade unions such as the International Workers of the World, commonly known as the "Wobblies" and led by the burly former hard-rock miner (and murder suspect) William "Big Bill" Haywood, had become increasingly avid in their use of force to obtain concessions from employers. Another small trade union, the International Association of Bridge, Structural, and Ornamental Ironworkers (an affiliate of Samuel Gomper's American Federation of Labor), set off at least twenty large explosions in 1908. That March, the ironworkers came close to blowing up New York's nearly completed Queensboro Bridge, the longest cantilevered span in the world.

Finally, of course, there were the *true* anarchists—the men and women who proudly called themselves by that label and proselytized the cause of overthrowing government, *all* government, by any means necessary. Generally portrayed as blood-crazed sociopaths or nihilists, most anarchists were actually rather dewy-eyed romantics. They believed, as the "Queen of the Anarchists" Emma Goldman put it, in the "beautiful ideal" of a just and peaceful society in which rules and conventions were abolished and all people lived in a libertarian Utopia. "I believe—indeed, I know—that whatever is fine and beautiful in the human expresses and asserts itself in spite of government, and not because of it," wrote Goldman in 1908. "Anarchists," she added, "are the only true advocates of peace." The rub was that, before the peace-loving Utopia could be attained, government and authority had to be dismantled, even if that dismantling required violence.

The actual number of self-professed anarchists in America—about twenty-five thousand—never represented more than a sliver of the country's population. Given the amount of ink they received from the press, it often seemed as if there were millions of them. Like communists and terrorists in later years, anarchists triggered enormous anxiety, and seldom more so than in the late winter and early spring of 1908. Over a six-week period, between the third week of February and the last week of March, a cluster of anarchist attacks sent jitters through the country. The first of these occurred on February 23 in Denver, Colorado, when an Italian immigrant named Giuseppe Alio killed a beloved Roman Catholic priest, Father Leo Heinrichs, during mass.

The second occurred in Chicago.

Chicago

The American racers drove into Chicago on February 25, a full day ahead of their nearest competitor, under a drizzle of cold rain and sleet. The Thomas Flyer was battered and filthy, and as sodden, according to the *Chicago Daily Tribune*, as "the deck of a ship in a hurricane."

The previous week had been grueling for both men and machines. After skirting the Great Lakes, the racers had entered the flatlands of Indiana and driven directly into one of the worst blizzards in the state's history. Snow fell to a height of four feet; drifts rose as high as seven feet. In sections, the racers' progress averaged one mile per hour, significantly slower than Frederick Cook's march to the North Pole. On one especially dismal day, the Thomas Flyer managed to travel just seven miles in thirteen hours. It was only horsepower that kept them moving forward at all—*real* horsepower, the kind supplied by the teams of draft horses on which they relied to pull them out of the hopeless drifts.

Altogether, the Thomas Flyer had managed to cover the thousand miles from New York to Chicago in thirteen days while traveling about twelve hours each day. The automobile, in other words, had taken about a hundred and fifty hours to make a trip the Twentieth Century Limited train service made daily in eighteen hours.

If the Americans' feat did not exactly qualify as a marvel of modern transit, the people of Chicago seemed not to notice. They were as enraptured as if the Thomas Flyer had plummeted from outer space. On the slushy gray afternoon the automobile entered the city, bonfires burned in Jackson Park to greet the racers. Fireworks exploded. Large crowds erupted in cheers along Michigan Avenue. When the car finally stopped

The Thomas car, stuck in Indiana snow, assisted by horsepower.

in front of the headquarters of the Automobile Association, men in bowler hats pressed around it like penguins huddling in a storm, eager to shake the hands of the crew before they were "rushed to hot baths and good clothes," in the words of the *Chicago Daily Tribune*.

The city the racers entered was, if possible, a more frenzied place than even New York. "The energy of New York is said to be mere leisure compared to the hustling of Chicago," wrote the British journalist and literary critic Charles Whibley in 1908, after visiting the city and finding it not to his liking. "Wherever you go," added Whibley, "you are conscious of the universal search after gold."

That search had produced a city of peculiarly American achievements. Chicago was the birthplace of the steel skyscraper, the nexus of the transcontinental railroad system, seat of the country's meatpacking industry, and, not least, gambling capital of the United States. "First in violence, deepest in dirt, lawless, unlovely, ill-smelling, irreverent," the great muckraking journalist Lincoln Steffens had written of Chicago in his 1904 book, *The Shame of the Cities*. Chicago was unencumbered by any of the old world pretensions that weighed upon Boston, Philadelphia, or New York. "The first impression of Chicago, and the last, is of an unfinished monstrosity," wrote Whibley, sniffing at the city with turned-up British nostrils. Part of what distinguished Chicago was that it could not have cared less what Charles Whibley thought.

Besides its other contributions to America, this city on the edge of the prairie was the breadbasket of American labor unrest. It was in Chicago that the most infamous labor riot in American history had occurred twenty-two years earlier, when a protest at Haymarket Square exploded with fatal results in 1886. Eight years after that, Chicago was at the center of another infamous episode of proletariat revolt, when Eugene Debs (who would be the Socialist candidate for president in 1908) led railroad workers on a nationwide strike against the Pullman Palace Car Company and shut down much of the nation's rail traffic.

So far 1908 had been a difficult year for Chicago's laborers. Even as immigrants poured into the city to partake of Chicago's rapid growth, that growth had been checked by the October panic. Immigrants kept coming but jobs were scarce, and now Chicago had one of the largest unemployed populations in the country. It simmered with poverty and resentment. The socialist "King of the Hobos," Dr. Ben Reitman, led the unemployed on angry marches through the city. Emma Goldman was expected in town any day to begin an extended speaking engagement, sure to rouse the rabble further. (During her visit, she and Reitman would begin an intense love affair.) Meanwhile, Chicago's three main gambling rings kept the city on edge with frequent attempts to bomb each other out of existence. The Chicago police force girded for the worst.

The arrival of the Thomas car, passing through on its whimsical journey around the earth, was a pleasant respite in this winter of discontent. Over the next three days, the crewmen received a heroes' welcome of banquets and speeches. On the morning of February 28, the Thomas Flyer, freshly polished and bedecked in flowers, took its leave of the city and headed west to cross the Mississippi River.

Three days later, on the bitterly cold Monday morning of March 2, a nineteen-year-old Russian-Jewish immigrant named Lazarus Averbuch boarded a streetcar on the west side of Chicago, where he shared a small apartment with his sister and several others. Averbuch's journey to the tonier north side took about an hour and a half. It was almost nine when he strolled up North Lincoln Place, a street of large houses in an affluent neighborhood. He came to number 31 and walked up the front

The Thomas car near Cedar Rapids, about March 1, deep in the infamous
Iowa "gumbo." The racers agreed that American roads were among
the worst in the world.

steps to ring the bell at the home of George Shippy, Chicago's chief of police. A moment later, the chief himself opened the door. Averbuch held out a letter.

Chief Shippy would later state that he'd had a premonition that a man would try to assassinate him. This may explain why he immediately grabbed Averbuch's arm and began wrestling him. What happened inside the front hall of Shippy's house after that is not entirely clear. The struggle between Averbuch and Shippy was soon joined by Shippy's wife, teenaged son, and driver. Shippy later claimed that Averbuch was armed with a gun and knife and fought, in the words of one newspaper, "with all the maddened fury of a fiend incarnate." When it was over, Averbuch lay dead on the floor riddled by six bullets, the chief's son was critically wounded by a bullet in the chest, and Chicago was in hysterics. That was the second anarchist attack of 1908.

The Plains

The morning Lazarus Averbuch visited Chief Shippy in Chicago, the Thomas Flyer was about three hundred miles to the west, making slow progress through the infamous dark mud of Iowa. A thaw had liquefied the roads into "black, slimy quagmires," wrote Montague Roberts, the American driver. The cloying "gumbo," as locals called it, kept the Thomas to a crawl of about seven miles an hour.

If road conditions were abysmal, worse than Roberts had ever seen, Iowa compensated with its population of friendly farmers. Always ready to lend a hand with directions or, if necessary, sacrifice a line of fence so the racers could cut across their fields, these men were the angels of the prairie, appearing when most desperately needed. "The Western people have won the first place in my heart on account of their sincere enthusiasm, their patriotism, and hospitality," wrote Roberts as they passed through Iowa.

Despite their sluggish progress, the American team still held a comfortable lead over its competitors. The Thomas was a few days ahead of the Italian Zust and the French De Dion. Neither the German Protos nor the other French car still in the race, the Motobloc, had reached Chicago. The third French car, the Sizaire-Naudin, was long gone, having dropped out back in New York. The Americans had dominated the race from the start.

The Europeans had a simple explanation for this. They believed the Americans were cheating. The crews of the French De Dion and the Italian Zust had filed a joint protest with the race committee accusing the Americans of several rule infractions, including borrowing horses to drag the Thomas through long stretches of Indiana snow. In fact, the Americans *had* used horses, more than once, but whether this constituted a breach of rules was unclear.

Roberts dismissed the protest as the carping of sore losers. He couldn't deny, though, that the American team enjoyed a home-field advantage. The same farmers who had been so provident to the Thomas team with horses and directions were often indisposed when the foreigners came knocking.

On March 4, the racers passed through Council Bluffs, at the western edge of Iowa, and drove over the Missouri River Bridge into Omaha, Nebraska. Here the excitement reached a pitch such as the racers had not experienced since Times Square. Eight cannons blasted to greet them. Factory whistles screeched. Thousands of Nebraskans—the largest crowd in Omaha's history, it was said—clamored around the mud-caked car, tossing flowers and waving flags.

To the racers, Omaha meant the treacherous Iowa mud, and nearly half the country, was behind them. They were probably too overwhelmed by exhaustion, cold, and attention to ponder the larger historical significance of the place they had reached. Distinguished mainly by stockyards and rail yards, Omaha did not look like much, but geographically and historically this small city between the prairie and the plains was a critical patch of American real estate. It was here on a bluff above the Missouri River, just to the north of downtown Omaha, that Lewis and Clark and the Corps of Discovery had first met western Indians (mainly from the Oto and Missouri clans) while on their way into the newly acquired lands of the Louisiana Purchase, and where they had informed the chiefs that a new "Great Father" now ruled them. Within a few decades of that historic parley, Omaha had become a jumping-off point for westbound settlers; and in the 1860s, the Union Pacific railroad had begun laying its track westward from Omaha, the start of the transcontinental railroad.

To white Americans, Omaha had been a stepping-stone into the west. To the Indians, it was more like a rock applied to the head. Whites poured across the west from Omaha, claiming land, killing the buffalo that had been the Indians' sustenance, introducing diseases the Indians had no immunity to and massacring tribes that fought back. By the time the Thomas Flyer arrived in Nebraska in March of 1908, the buffalo were long extinct and the Indians of the Plains reduced to abject poverty.

Now and then the warriors still put on their feathered headdresses, mounted their horses, and looked for all the world like the fierce Indians of western legend, but they did this only for show now. Mainly, they did it to humor the white photographer Edward S. Curtis, who was wandering the northern plains in 1908, using his camera to capture the Indians as they once had been. Curtis was shooting photographs that would become among the most iconic ever made of the American west, beautiful, haunting images with titles like "A Gathering War Party" and "Winter Hunter." These were largely re-creations when Curtis took them. The wars were all over. The only large game to hunt now were the mangy cattle provided by the Bureau of Indian Affairs to keep the Indians from starvation.

The American racers did not see any buffalo, but they did meet William Cody in Omaha. Better known as Buffalo Bill, Cody was in Nebraska (where he owned a ranch) on the day of their arrival and

"Ready for the Charge"
by Edward S. Curtis, 1908

greeted them on behalf of Omaha's official reception committee. Years earlier, Cody had been a practicing frontiersman and buffalo-hunter, but now, at fifty-seven, he was better known as the impresario of "Buffalo Bill's Wild West." Cody's show traveled the country like a circus, from small cities to major venues like Madison Square Garden in New York (where it would play in April 1908). Featuring "Real Red Men of the Plains," along with an assortment of "Wild-West Girls," and cowboys "Direct from Ranch and Prairie Ranges," Cody's popular show promised an authentic western experience. And to a certain extent, it delivered one; Cody hired real Indians and real cowboys, not costumed actors. It's just that the Wild West represented on Cody's stage, like Curtis's photographs, had little bearing to reality in 1908.

The mythology of the Wild West died hard. In some ways, this mythology was more seminal to the identity of Americans than the heroics of the Revolution or the Civil War. The epic of *How the West Was Won*—how savage lands, wild Indians, and cowboy outlaws were brought to heel—shaped not only westerners' identity of themselves, but easterners' identity, too. The irony of the Wild West by 1908, but also the moral of it, was that, having been won, it was less wild—less complicated, less savage, less dangerous—than the east. There were still occasional flashbacks (as occurred in New Mexico in late February when legendary sheriff Pat Garrett was shot down on a dusty road), but compared to New York, or Chicago, or a hundred other places east of the Mississippi, the far west was a land of tranquility.

On March 5, the crew of the Thomas drove out of Omaha and headed west to Cheyenne, Wyoming, through fields of wheat and cattle pastures. "The roads now make one believe we are surely in the Wild West," wrote Montague Roberts. "At times we would drive for miles and only see once in a while a farmhouse or a cattle ranch, possibly a cowboy on horseback. We met one cowboy who gave us a race for two miles and lost out, but he took it in good spirit and said his horse was tired."

Two hundred miles out of Omaha, they came to the town of Kearney, Nebraska. When the racers stopped there in late afternoon to post telegrams from the local Western Union office, they cordoned off the Thomas Flyer with a rope, as if it were a traveling museum piece. The rope was to protect the car from overzealous crowds who evidently felt a need to leave their mark on it. People scratched their names into the car's paint, stole pieces of it as souvenirs, or tied handkerchiefs to it in the fond hope their bits of cloth would end up in Paris. (A false hope; the crewmen were short of handkerchiefs and quickly appropriated them.) They peppered the men with questions. Was that mud on the car the infamous gumbo of Iowa? Were the racers really going to drive across the Bering Strait? What was the latest news on the protest from the Europeans? (Tired of answering that last question, Montague Roberts pretended to mishear the word "protest" and told his interrogator that the Protos was still back in Chicago and doing fine.)

The racers may have become slightly exasperated after several weeks of questions, but they could hardly blame the Nebraskans for asking. To the people of the plains, this vessel that had come from Times Square—where a giant electric ball fell from the sky and women smoked on New Year's Eve—and was on its way to Siberia and Paris was an object of intense fascination. All the more fascinating because the car's presence was, for most towns, only fleeting. The Thomas stopped fifteen minutes in Kearney, then was off again in pursuit of the falling sun.

The racers followed the Platte River through western Nebraska, just as the first white settlers had done. They ran alongside the old Oregon Trail, parallel to the tracks of the Union Pacific and beneath the east-west telegraph lines. Following these threads that had bound east to west a generation or two earlier, the Thomas payed out a new and invisible thread between east and west, and between past and future. The new west would not be settled by men and women in prairie schooners or on horseback. It would be settled by people in automobiles.

Lions and Lambs

Two weeks before the season's official due date, the first sign of spring arrived in New York City. The air was still crisp on the afternoon of March 10 but the ice on the Hudson River had begun to melt and fracture. New Yorkers out enjoying a splendid Tuesday in Riverside Park suddenly heard shouts from the river. Looking over the water, they saw three boys floating away on a pan of ice.

The boys had been playing tag on the ice when it broke from the shore and drifted out into the swift current. As they traveled down the river, passing Eighty-ninth Street toward New York Harbor, the boys appeared not the least bit troubled by their predicament. They danced about on the ice mirthfully, waving their caps at the people on the shore who called to them. Several concerned bystanders attempted to launch a boat to rescue the boys but were frustrated by the ice clumped near the boathouse. Before anything could be done, a passing tugboat raised a rough wake on the water. The pan of ice heaved up, seesawed, and broke into two pieces, separating one of the boys from his mates. The three boys were no longer mirthful.

Just as matters were looking desperate, something strange and unexpected happened. Half a mile from their starting point, the two ice pans and their respective occupants slipped out of the downstream current and drifted back to shore. Some men managed to throw a rope to the boys and pull them in closer. A policeman reached out and plucked them to safety.

All three boys vanished before the policeman could take their names, but an enterprising *Times* reporter later tracked one down to his home on West Ninety-ninth Street. The boy's name was John McCarthy and he was fourteen. At first, the boy denied any knowledge of the escapade on the river; under interrogation by the reporter and his suspicious older sister, though, he finally confessed his role in it. His sister assured him he'd receive a sound beating when their father retuned home and heard the news. "The carpet cleaner handle for yours," she said.

"I don't care," responded the boy with an insouciant shrug. "I had a bully good ride."

Teddy Roosevelt could not have put it better himself.

A journalist named William Bayard Hale visited the president later in the month. He came to gather material for a long profile of Roosevelt— "A Week in the White House"—later published in the *New York Times* (then again as a book). As Hale arrived to begin the week, the capital was experiencing its first effulgence of spring. The days were warm, the trees budding, the flowers starting to bloom. Outside on the White House lawn, the president's youngest son, ten-year-old Quentin, had marked out a diamond to play baseball with his school chums. Inside the White House, Hale found the president in high spirits, his office "filled with consistent good-humor, breaking literally every five minutes into a roar of laughter."

Despite the good cheer, the president had serious matters weighing on his mind. For one thing, he had just accepted an invitation from

Japan to send the Great White Fleet to Tokyo. He could not easily have turned the mikado down—the insult would have been colossal—but by accepting, he had committed the greater part of the U.S. Navy to what were, potentially, the least friendly waters on the planet. Whether Japan's invitation was a magnanimous attempt to repair soured relations between the two countries or a piece of cheese on a monstrous trap only time would now tell. "It is recognized," reported the *New York Times* on March 21, "that there is a considerable element of risk."

Also commanding the president's attention was the upcoming 1908 election. Roosevelt had pledged not to run for a third term. Instead, he had taken the unusual step of naming his successor—William Howard Taft, his secretary of war—and dedicated himself to getting Taft elected.

On Thursday, March 26, Roosevelt met with Taft, an obese and amiable fellow, and one of Roosevelt's dearest friends. (Their sons were friends, too; little Charlie Taft played on Quentin's baseball team.) Many pundits remained convinced that the president, for all his loyalties to Taft, would throw his hat back into the ring and walk away with the election himself. But they were wrong. As much as Roosevelt would have liked to run again, he had committed himself to staying out of the 1908 election, and he did not intend to go back on his word. He felt confident that Taft was the man to carry on his program and ensure his legacy. Before he could go in peace, however, he needed to make sure Taft locked the nomination at the upcoming Republican National Convention in June.

Not the least of Roosevelt's preoccupations that week in late March were the recent anarchist attacks in Denver and Chicago. The press was predicting a seasonal recrudescence. "The warm days of spring," *Broadway* magazine warned, "breed violence in the blood." As Congress debated new restrictions on immigration to keep those with anarchist tendencies out of the country, the president ordered a crackdown on anarchist sympathizers already inside the borders. On Monday, March 23—the first day of Hale's week with the president—the *New York Times* reported that Roosevelt had personally directed his postmaster general to ban an anarchist periodical, *La Questione Sociale*, from the mails. "They are, of course, the enemies of mankind," the president wrote of anarchists, "and every effort should be made to hold them accountable for an offense far more infamous than that of ordinary murder."

Acts of violence not specifically attributable to anarchists but anarchic nonetheless punctuated these last days of March. Over the weekend,

Night Riders killed a tobacco farmer in Carlisle, Kentucky. On Monday, March 23, dynamite exploded inside a newspaper plant in Rock Island, Illinois. On Wednesday, a bomb blew up a mill in Hampton Junction, New Jersey, and another bomb blew up a steel railway bridge near Perth Amboy. On Thursday, the Black Hand dynamited a bank on Elizabeth Street in New York City. And on Friday, March 27, in Washington, D.C.—just a few blocks down Pennsylvania Avenue from where the president was conducting business under Hale's observation—Congressman James T. Heflin of Alabama, a vocal advocate of Jim Crow laws, pulled out a pistol and shot a black man on a streetcar. "The negro," reported the *New York Times*, "had insulted him."

The main piece of intelligence Hale brought out of his extended visit to the White House was that the president was himself no rabble-rouser—in no way "a dipsomaniac, a cigarette fiend, a victim of pathological ego enlargement." He was, rather, a sane, rational man who lived in accord with the best practices of scientific management. Hale marveled at Roosevelt's ability to negotiate his demanding schedule, fluidly shuttling visitors in and out of his office, shuffling from subject to subject with ease and mastery. "His mind is orderly; its contents are thoroughly arranged; his workshop is scrupulously neat," wrote Hale. "The division and subdivision of his day is the perfection of system." Considering how convincingly the president's detractors had portrayed him as a raving lunatic, this must have been welcome news to his constituents.

The last day Hale spent with the president, Saturday, March 28, was wet and muggy. Despite the enervating weather, the president met over two hundred visitors, radiating warmth and lucidity before each. The only disruption to his perpetual turnstile of meetings occurred when a man who had recently returned from a lion hunt in Africa stopped by at the White House. Roosevelt put aside all other business and questioned the man for half an hour. He had not yet made any firm decision about his postpresidential plans, but the lions of Africa were beckoning him.

As the president sat in his office and spoke of lions and African safaris, the warm drizzle continued to fall outside, over Quentin's baseball diamond and the new blooms in the garden. March was going out like a wet lamb. But this was a lamb, before the day was done, who roared.

Union Square

At three-thirty that afternoon, in New York City, a young man entered Union Square from the west. Nobody took much notice of him because he looked like many of the other thousands of young men who had been loitering and shouting in the square all afternoon. Most of them were gone now, but a few, like this young bearded man, remained. His name was Selig Silverstein. He was a slight twenty-two-year-old Russian immigrant, "scarcely more than a boy," according to the next day's *New York Tribune*, "with the stunted body and anemic appearance of the slum dweller and sweatshop worker." A cigarette drooped from his mouth. The drizzle that had fallen over the city for much of the day had stopped. The paving stones remained wet and glistening and littered with remnants—cigar and cigarette stubs, leaflets and circulars—of the crowd that had gathered there earlier in the afternoon to take part in a demonstration sponsored by the Socialist Conference of Unemployed.

That demonstration never occurred. In light of recent anarchist activ-

Union Square in New York on Saturday, March 28. Driven from the park, a crowd mills around the nearby streets.

ity, and with memories of Denver and Chicago still fresh, the Parks Department had decided at the last hour to ban it. When crowds began arriving by the scores, they found the park roped off and occupied by baton-wielding police who would not let them enter. Having nowhere else to go, the demonstrators clogged the streets around Union Square. Their presence gave the area the appearance, the *New York Tribune* reported, "suggestive of some Russian city rather than New York." One of the men who had intended to speak at the demonstration, a well-to-do socialist named Robert Hunter, later drew a similar comparison; but for him, it was not the crowd that brought Russia to mind, it was the police. "As a very considerable portion of the crowd approached Seventeenth Street and Union Square East a score of police on foot and as many on horseback charged the crowd with incredible brutality," recalled Hunter to the *New York Times*. "A few of them seemed as fanatical and fiendish as Cossacks."

Another of the demonstration's organizers, a sculptor named Bruno Zimm, approached Inspector Max Schmittberger, a famously tough high-ranking police veteran. Zimm pulled a book from his pocket and waved it at Schmittberger. "This is the constitution of the United States, and it is on this that we demand the right of free speech." Inspector Schmittberger responded by yanking his nightstick from his belt. "This is bigger than the constitution right now," he growled. "Now, move on."

But all those confrontations were well in the past when Selig Silverstein entered Union Square at three-thirty. The square was quiet now. The demonstrators had withdrawn, many going back home, others repairing to the nearby Café Hungaria or loitering on the perimeter of Union Square. The police were starting to allow pedestrians back into the park.

Later, witnesses would recall seeing a parcel in Selig Silverstein's hand about the size and shape of a grapefruit. They would remember how he paused when he neared a group of policemen mustered by the statue in the center of the park. He fumbled for a moment with his parcel. A flare of sparks shot up. He lifted his arm to throw the object—but too late. It exploded in his hand with a reverberation, according to the *Times*, "like the discharge of one of the great guns of Fort Wadsworth." Men and women a block away fell to their knees. Windows and doors rattled for blocks around the square. For a moment, all was silent, as a plume of thick blue smoke rose up over the fountain into the budding trees. And then—pandemonium. The demonstrators who had been lingering around the park began rushing toward the explosion, or away from it, and police began swatting people with nightsticks. Newspaper reporters who had been milling around the park all afternoon ran to the spot near the fountain where two men lay on the ground, blood pooling around their bodies. The second man would immediately be suspected as a collaborator of Silverstein's; in fact he was an innocent bystander in the wrong place at the wrong time, not an anarchist at all (his distraught wife would later explain) but a good Republican.

Silverstein, incredibly, was alive despite severe mutilation to his right arm and the right side of his face. A quick-thinking policeman used the cord of his nightstick to tie a tourniquet around Silverstein's arm to prevent him from bleeding to death. Silverstein mumbled something, first in English, then Yiddish and German. A German-speaking policeman was brought over to translate. "Yes I made the bomb, and I came to the park to kill the police with it," said Silverstein. "I am very sorry that I did not make good. As for my life, that is nothing." He fashioned the bomb, he would tell the police, by stuffing nails and nitroglycerine into the bulb

Twenty seconds after the explosion in Union Square Park, Selig Silverstein and an innocent bystander lay on the ground.

of a brass bedstead. He'd tried to light it with his cigarette, but in his nervousness and haste he'd made a critical error. "I put the cigarette in the wrong hole."

Before he died, Selig Silverstein insisted that he'd acted alone, but the police and the newspapers assumed otherwise. Rumors swept across the city that a hundred more like him were at large and lying in wait. For several days, as police pursued anarchist collaborators, the press provided a steady thrum of ominous reports: four men had been seen accompanying Silverstein to Union Square; letters from the anarchists Alexander Berkman and Emma Goldman had been discovered among his possessions; more anarchy was planned; greater death plots were in the works. All of which turned out to be untrue. The dreary fact was that Selig Silverstein, like Lazarus Averbuch, had acted alone, driven by his own demons and not by Emma Goldman or anyone else.

And yet it was impossible not to view Union Square as another front in the class war that had been tearing at the country for years. America seemed, at times, a blundering schizophrenic caught between opposing impulses. Gluttonous capital or violent labor? Feckless wealth or heedless revolution? Oligarchy or anarchy? Neither position had much to recommend it. The working-class agitators were more explicitly violent, but the rich were far from blameless. Even the *New York Times*, no friend of the proletariat, suggested the rich shared responsibility for the sort of "outrage" that Selig Silverstein perpetrated in Union Square. "The disregard of ordinary prudence in the conduct of their domestic relations, the willful neglect of the proprieties, among rich people," declared the *Times* in an editorial a few days after the bombing, "tend to increase the volubility of the agitators against existing social conditions." Both sides, rich and poor, were engaged in a symbiotic, dysfunctional, self-centered, and highly self-destructive relationship. Both operated, as another critic put it, under "jungle law."

There had to be a better way.

In fact, there was.

CHAPTER FIVE

Ultima Thule

MARCH 31–APRIL 21

Standing on this spot, I felt that I, a human being, with all of humanity's frailties, had conquered cold, evaded famine, endured an inhuman battling with a rigorous, infuriated Nature in a soul-racking, body-sapping journey such as no man perhaps had ever made.

—Frederick Cook

Shortly before seven o'clock on the last evening of March, as police continued to scour the city for Selig Silverstein's collaborators, five hundred prominent men and women gathered in the Grand Ballroom of the Waldorf-Astoria on Fifth Avenue. The occasion was a formal dinner on behalf of the National Playground Association. The guest list included an array of college presidents, magazine publishers, department store owners, politicians, and millionaires. Anne Morgan, daughter of J. P. Morgan Sr.—and sister of Jack—was among the guests, as were John D. Rockefeller Jr. and his wife. William K. Vanderbilt and his estranged wife, Alva, were there, too; they came to hear their daughter, Consuelo—better known as the duchess of Marlborough—deliver the keynote speech of the evening.

Shortly after ten o'clock, as dinner was winding down, the gentlemen were given permission to smoke "by the gracious consent of the ladies." Cigars and cigarettes appeared from waistcoat pockets. Flames ignited, filling the room with flickers of light. Smoke curled toward the gilded ceiling.

The duchess of Marlborough rose and walked to the dais. She wore a shimmering gown of blue satin and gold lace. A band of diamonds clasped her long neck and strings of pearls draped over her shoulders. With her small mouth, her wistful eyes and pert little nose—features John Singer Sargent had captured in his portrait of the Marlboroughs— she was still delicately pretty and girlishly slender at thirty-one, despite two children and a loveless marriage to the ninth duke of Marlborough.

The duchess had come to the Waldorf this evening to speak on "The Responsibility of Women." Standing at the dais, reading from her text, she delivered what amounted to a pep talk, exhorting American women to devote themselves more diligently to the improvement of social ills. "Having had the opportunity of coming in touch with the public work being done by women in England," she told the audience, "I hoped I might be able to sound a note of encouragement to my own country-women who are devoting their lives to bettering the social conditions of the poor in this city."

In their postprandial stupors, the guests might be excused if they did not listen too attentively. The duchess's words were less important than the fact of her titled presence. She may have been a sad cliché—a beautiful spoiled heiress trapped in a doomed international marriage—but she was indisputably a duchess, and beautiful, and very rich. Even the press that disdained international marriages as unpatriotic and unworthy of American women could not help being smitten. Tomorrow, nearly every New York newspaper would feature the duchess's speech on its front page.

After Consuelo took her seat, the guest of honor, a well-regarded English novelist, Mrs. Humphrey Ward, delivered a speech comparing good works in England to good works in America. Then Jane Addams stood.

Jane Addams provided a stark contrast to the duchess. She lacked money, title, and beauty, and she dressed plainly and simply, like the frumpy spinster of forty-eight she was. But as the founder of Hull House in Chicago's impoverished Nineteenth Ward, the first so-called settlement house in the United States, Addams had devoted her life to carrying out in deeds the sort of charitable acts that Consuelo Vanderbilt recommended in words. She had, by 1908, achieved her own kind of nobility, albeit of a very different sort than the duchess's.

As an orator, Jane Addams eschewed the flowery rhetoric common at these sorts of functions. She was matter-of-fact and plainspoken—plain

Jane to the bone. Her words, like Consuelo's, meant less this evening than her presence: her gravitas, her renowned goodness, her renouncement of worldly things. She championed the poor, but she did not threaten the rich. She spoke to them gently, as a friend and compatriot, and asked them to recall their better selves. In Jane Addams, the frumpy matron of Hull House, the millionaires from the mansions glimpsed their conscience, floating in the distance through a haze of tobacco smoke.

There are many terms that were used to describe Jane Addams during her lifetime. She was a settlement house worker, a reformer, a social worker, a pacifist. But by 1908, women and men like Jane Addams—which is to say, Americans who devoted themselves to bettering American society—had begun to coalesce into a movement known as progressivism. Actually, the term did not come into general use until a few years later; but the movement was already well under way.

Progressivism: the very word is enough to induce yawns in high school history students. They were not, on the face of it, a terribly exciting bunch, the progressives. Largely middle-class, middle-aged, and middle-of-the-road, they paled as personalities beside the extravagant rich and the violent poor. While political figures like Teddy Roosevelt and Senator Robert "Fighting Bob" La Follette later marched under the progressive banner, most progressives were nothing like these larger-than-life men. Rather, they were mild-mannered, well-meaning do-gooders. They were psychologists and lawyers, ministers and educators, doctors and journalists. They were men like Herbert Croly, an obscure and uncharismatic thirty-nine-year-old magazine editor who was so crippled by shyness that encounters with him were often, in the words of writer Edmund Wilson, "baffling." In his off-hours that year, Croly was composing the third draft of a dense but remarkable book he would entitle *The Promise of American Life*. Published in 1909, the book would sell no more than seventy-five hundred copies during Croly's lifetime—a suitably mid-list figure for a middle-of-the-road progressive—but it would contain a powerful, sweeping, and highly influential vision of America as a place where the riches produced by unfettered free-market capitalism met with the democratic virtues of compassion and fairness, combined with an all-American confidence that even the most intractable social problems could be fixed.

Originally, the agenda of progressives focused on a handful of priorities, mainly cleaning up corrupt city governments and curbing the abuses of monopolies. From these narrow objectives, though, progressives had turned their attention ever more compulsively to a multitude of social ills. It was as if, having cracked open the door on a messy closet, they could not stop cleaning until they scrubbed every crevice in the house. They pushed for new labor laws, new railroad regulations, new rules to ensure a cleaner and safer food supply. They sought better housing for the poor and better public schools. Nature conservation, disease eradication, women's suffrage, temperance—all were reform movements that gained ground in the early twentieth century under the general outlines of progressivism.

So broad was the progressive agenda by 1908 that it is almost absurd to speak of it as a coherent movement. But several basic assumptions bound the reformers. The first of these was that Americans had a duty to improve the moral and physical quality of their fellow Americans' lives— a clear rejection of laissez-faire principles that had dominated political thought of the late nineteenth century. A second shared assumption, drawn from the first, was that local and national government had a substantial role to play in carrying through on this obligation. The third assumption was that Americans, transcendentalists all, were endowed with a singular capacity to transform and uplift themselves.

It was the last assumption that explained what five hundred prominent men and women were doing gathered in the Waldorf-Astoria on that evening at the end of March. They came to honor playgrounds: play-

Jane Addams, circa 1908

grounds not simply as places where kids played, but as places where progressives exercised their faith in the capacity of humans to better themselves. Progressives were convinced that the grim lives of underprivileged children—gantlets of disease, crowded tenements, and unwholesome activities—could be improved by placing them in conditions conducive to healthy physical and mental development. This may be an obvious insight to Americans of the early twenty-first century, but to Americans of the early twentieth century, when social Darwinism still held sway, it was a revelation.

Social Darwinists believed that nature, not nurture, determined how children grew and behaved. To a social Darwinist, mental development and moral character were essentially predetermined hereditary characteristics. Although social Darwinism had begun to lose its sway in some quarters by 1908, notions of biological determinism—as we have seen in the Thaw trial—remained a powerful influence on American thought. Insanity, criminality, anarchy—all were considered forms of genetic degeneracy; all demonstrated unfitness in the great survival sweepstakes. The fact that these so often showed up in certain impoverished ethnic groups, especially recently immigrated Eastern Europeans and southern Italians, suggested to social Darwinists not that low economic status caused people to behave badly, but that low economic status was another indicator of biological inferiority. Granted, wealth was no guarantee of sound mental health in this view—see Thaw again—but poverty indicated a strong predisposition to all sorts of degenerate behavior.

What this meant for children—poor children, specifically—is that some were *born* bad. Criminality was coded in their blood, in the pigment of their skin, in the very bone structure of their bodies. Indeed, it was so molded into their physiognomies that a trained eye could detect it on sight. In his 1908 book, *Young Malefactor: Juvenile Delinquency, Its Causes and Treatment,* a criminologist named Thomas Travis described a few of the usual *stigmata,* as criminologists referred to physiological anomalies: "[A]bnormally shaped head-bones ... abnormal palates, sinister faces, deflected noses, prognathous jaws, and general cranial asymmetry."

Another criminologist, William McDonald, went before Congress in March of 1908 to urge federal funding for a research laboratory devoted to studying the physiology of the "weakling classes." Along with his testimony, McDonald included descriptions and drawings of a great many "scientific instruments of calculation" that would precisely measure physical defects beyond what the eye could see. A "psychograph" measured trembling; a "glosso-dynamometer" measured the strength of the tongue to resist pressure; a "laryngograph" measured the movement of the larynx during speech. One especially alarming looking device, the "palatograph," consisted of a band worn tight around the subject's head and connected to a small metal wire shaped like a miniature tennis racquet. The looped wire fit in the subject's mouth like a horse's bit and was used to record the movement of his palate during speech.

Many criminologists considered defectives beyond repair; the best thing was to send them to prison or a remote island. Thomas Travis was a little more optimistic. He was convinced that some physical defects correlated to delinquency could be ameliorated by surgery. Fix the exterior and you could fix the person beneath. "The results are in many cases simply amazing," wrote Travis of such surgeries.

> The writer has seen juveniles he would have suspected of natural backwardness or criminality—children repulsive almost in their mal-development and certainly ugly in their disposition; he has seen such a change wrought in six months as a result of the operation that the children were scarcely recognizable. The faces were illuminated with intelligence, the body erect with health, and the mind and morals wonderfully improved.

Progressives saw things differently. Rather than surgically modify the child, better to modify his environment, for it was how he lived, not how he was born, that mattered most to his development. "Heredity determines what each individual child may become; environment determines what he does become," is how the physician Dr. Carl Kelsey put it in a 1908 lecture at New York's School of Philanthropy, defying the conventional wisdom. "The difference between children at birth is far less than we have been accustomed to think."

At a time when most poor children played among the filth and perils of the streets, the Playground Association, founded in 1906, was devoted to the proposition that intelligently designed playgrounds could shape children into healthier, more wholesome citizens. In state-of-the-art facilities, boys and girls would play on equipment designed to promote their physical well-being. Their moral well-being, meanwhile, would be supervised by a full-time playground attendant who would direct their play into age- and sex-appropriate behavior and keep them out of harm's way.

In retrospect, the idea that a child could find redemption on a swing set might sound as preposterous as the assumption that a child's character was set in his or her bone structure. But to playground advocates like Jane Addams, the chance for a child to play in a safe, clean, supervised environment was nothing less than a chance to change his or her destiny. Conversely, Addams wrote, "To fail to provide for the recreation of youth, is not only to deprive all of them of their natural form of expression, but is certain to subject some of them to the overwhelming temptation of illicit and soul-destroying pleasure."

With that last unfortunate phrase—"temptation of illicit and soul-destroying pleasure"—Addams highlighted the fundamental problem with progressives. A tone of priggishness and sanctimony seeped into almost everything they said and did. When Jane Addams wrote of soul-destroying pleasure, her readers knew what she meant: sex.

Even by the squeamish standards of the white middle class, progressives were notably troubled by sex. They rejected the social Darwinists' fixation on hereditary determinism but held on to Victorians' repugnance for all things primitive or base. Sexual lust was exactly the sort of nature they hoped nurture would vanquish.

Ironically, despite the forward-sounding name by which we know them, it was progressives who worried more than anyone about progress. They were forever deploring the degraded moral culture of early twentieth-century America, the "relaxation of the moral muscle," as Percival Chubb, president of the Society for Ethical Culture, put it. They seemed to want nothing more than to turn back the clock to a more innocent and sexless time, when the world behaved more like—well, more like them and less like the bawdy poor and the lascivious rich. At times, they sounded not so much progressive or even Victorian as puritanical. They were the last people on earth who should have been telling children how to *play*.

Progressives are easy to belittle nearly a century after their extinction. Who can be more insipid, more aggravating than a straitlaced do-gooder who insists that everyone behave as he does? The fact that most of what progressives proposed never came to fruition (with the notable exception of Prohibition) adds to their efforts a whiff of ineffectuality and dilettantism.

In the end, though, for all their faults, they were a praiseworthy, even heroic, group of Americans. Not for what they achieved, but for what they hoped to achieve. Unlike anarchists or socialists, they never advocated wholesale overthrow of the existing social order. Rather, in their unexciting way, they attempted to remedy the vast disparities between the wealthy and the poor, and to do it with compassion and fairness. Encouraging the poor to have faith in the promise of American life, while encouraging the rich to honor this faith, they charted a middle road into the wilds of the twentieth century.

The Ice World

Had he chosen a different path in life—had he remained a general practitioner and family man in Brooklyn, for instance—Dr. Frederick Cook would have made a fine progressive. The profile fit him comfortably. He was a man of firmly middle-class background, modest, moderate, and fair-minded. His conflicted attitude toward the Eskimos demonstrated a view of the "savages," as he called them, that was at once sympathetic and condescending, and entirely characteristic of the progressive mindset. (Though, to his credit, he was less judgmental than a typical progressive.) Frederick Cook was exactly the sort of man who would have admired a scientifically designed playground for disadvantaged children.

But he had another fate in store. Whatever immoderate urge it is that makes a person want to leave "the jungle of city life," as Cook called it, and dwell in the "ice world" possessed him and he could not shake it. Once experienced, Cook wrote, "the lure of the Arctic becomes a permanent drawing power for life." And so, on that same night Jane Addams stood on the dais in the Waldorf looking back at her audience through a scrim of tobacco smoke, Frederick Cook was hundreds of miles out on the ice of the Arctic Ocean peering through a mist. Earlier in the day, Cook believed he had seen land in the distant west—he was sure of it—a long hump rising from the plain of ice. He named the hump Bradley Land in honor of John Bradley, the man who had funded his expedition. Then the mist closed in and the land vanished as if it had never existed.

* * *

Cook and his party of ten men had departed Annoatok nearly six weeks earlier. They had walked across the ice of Smith Sound, then traversed the width of Ellesmere Island before turning northwest at Nansen Sound. Advancing as much as twenty-five miles a day over snow-covered land or on ice adjacent to the east coast of Axel Heiberg Island, the expedition had gone more or less as planned. The caches of wild game for which Cook had arranged months earlier had provided the men with fresh meat. Even fresher meat came from the musk ox and bear they hunted along the way. In the evenings, when temperatures fell as low as minus seventy-nine degrees Fahrenheit, Cook and his men warmed their igloos with oil lamps and drank scorching coffee laced with sugar, then woke in the mornings to the cheer of the sun rising a little higher each day.

By the middle of March, they had come to the end of land—*real* land. At the black cliff face of Cape Svartevoeg, at the edge of the Arctic Ocean, some five hundred and twenty miles south of the Pole, Cook had bid farewell to most of the men. He almost immediately reduced his party further, to just three men, Etukishook, Ahwelah, and himself. They took with them two sledges loaded with the pemmican and other supplies. In the manner of most polar explorers, they walked alongside or behind the sledges rather than ride on them. Between the cold and the exertion of walking, they burned quickly through the pemmican. They could expect no more fresh meat, with the possible exception of the dogs they would eat if necessary. Cook had chosen twenty-six of the finest animals to draw the sledges. One by one, as per Cook's plan, the weaker would be slaughtered to feed the stronger, until, at last, only the six strongest would survive to haul back what was left of Cook's supplies.

Averaging about fifteen miles a day, Cook and the two Eskimos walked over the undulating and drifting ice field, scaling pressure ridges where the ice pushed up in ragged heaps, and crossing over new ice that was thin, yellow, and smoking. One night, while Cook slept, the ice cracked and opened beneath him. Cook woke to find himself falling into the ocean. A moment later, still tucked into his sleeping bag, he was thrashing in the frigid water. Only the quick actions of Etukishook and Ahwelah saved him from death.

After that, Cook had begun to see things in the ice field, fabulous and strange mirages produced by cold air and deflected light. "Out of the pearly mist rose marvelous cities with fairy-like castles; in the color shot clouds waved golden and rose and crimson pennants from pinnacles and domes of mosaic-colored splendor. Huge creatures, misshapen and grotesque, writhed along the horizon and performed amusing antics." On March 30, that great hump of land to the west appeared: *John Bradley Land.* Cook estimated it to be an island about a thousand feet high and fifty miles long.

In fact, there was—there is—no land where he saw it.

Cook and the two Eskimos now advanced northward in a delirium of exhaustion and deprivation. They had walked more than three hundred miles from Cape Svartevoeg; they had another two hundred miles to go before they reached the Pole. The small comforts they had enjoyed earlier in the journey were now absent. The topography was relentlessly the same, the days grueling duplications of slogs over a rough surface of crevasses and pressure ridges. Earlier, they had been kept company by animal trails, even by occasional sightings of animals. Now, all visible forms of life, other than their own, vanished. They were beyond life, traversing "a blank space of the earth," as Cook called it. "Beyond the eighty-third parallel," he added, "life is devoid of any pleasure."

At the end of each day, the men built a quick igloo, fed the dogs, then crawled into sleeping bags to gnaw on their daily ration of pemmican. The Eskimos promptly fell asleep after eating, but for Cook sleep became impossible. Exhausted as he was, he spent the hours lying in his sleeping bag as images and thoughts swirled through his brain in "a delirium of anxiety and longing."

In his writing, Cook does not dwell on the insomnia that would torment him the rest of the way to the Pole, but it must have been appalling. Through the walls of the hastily made igloo he would hear the wind whistling and the dogs howling miserably. The ten-foot-thick ice beneath him would crack and shift, drifting to the southeast on the currents of the Arctic Ocean. Below the ice, the ocean dropped to a depth of nearly a mile. At any moment, as he knew from experience, the ice could open and swallow him. No wonder his thoughts scurried to dark places. "Lying wakeful in that barren world, with my companions asleep, I felt what few men of cities, perhaps, ever feel—the tragic isolation of the human soul—a thing which, dwelt upon, must mean madness."

He did seem, in fact, to go a little mad. The mirages that had accompanied him since he and his companions took to the ice now began to assume more haunting "phantom" and "wraith-like shapes." Cook heard voices in the wind and felt "unseen forces" tugging at him, urging him to keep moving forward despite his overwhelming fatigue. "All Nature exulted in a wave of hysteria," he later wrote. "We moved in a world of delusions."

The delusions were fueled by sleeplessness and malnutrition. But it probably didn't help that the usual signposts of space and time had dissolved. Open whiteness reached out in all directions. The very idea of space—of *place*—was confused by the constant movement of the ice. The three men were traversing a kind of conveyor belt; even as they walked due north, they drifted constantly east. Further confusing their already boggled minds, the sun now circled them day and night, rendering day and night meaningless concepts. Time itself was meaningless at the top of the earth, where all the lines of latitude, and all the time zones of the earth, converged.

Cook and his men were walking through a placeless place and a timeless time. The only intelligible cadences were their feet crunching the ice, their rasping breath, their hunger and exhaustion. At the top of the world, the world turned upside down, topsy-turvy, inside out. "In our dreams Heaven was hot, the other place was cold," wrote Cook. "All Nature was false; we seemed to be nearing the chilled flame of a new Hades."

The End of the Road

The topsy-turvy feeling Frederick Cook suffered near the top of the world was psychological. For the crew of the Thomas Flyer, it was more palpable. During the same days of early April that Cook was trudging across the ice of the Arctic Ocean, the American racers, who had driven over land almost four thousand miles from the edge of the Atlantic, now found themselves huddled in a small boat on the storm-tossed Pacific.

For nearly a week, the Thomas Flyer and its crew had been lost off the west coast of North America. Last seen boarding the steamship *Santa Clara* in Seattle, en route to the port of Valdez, Alaska, men and machine had steamed out of Puget Sound and directly into the path of a pounding storm. The little steamer with the automobile lashed to its deck pitched and brawled toward the Bering Sea. Snow fell and waves crashed over its decks, spraying the Flyer with corrosive needles of salt water. When carrier pigeons were released from the boat to take news of the hard voyage back to land, the birds had no sooner reeled into the sky to orient themselves than a flock of seagulls dove in and killed every one.

Meanwhile, the rough-hewn little town of Valdez awaited the automobile with growing anticipation. The expected day of arrival passed, then another. Finally, on the afternoon of April 8, 1908, two days behind schedule and seven days out of Seattle, the *Santa Clara* appeared chuffing out of Prince William Sound into the fjord of the port of Valdez. The eight hundred residents of Valdez gathered along the shore to watch the automobile approach.

The fact that the Thomas Flyer was arriving lashed to the deck of a boat, and not on its own four wheels, did little to diminish the excitement in Valdez. Few Alaskans had ever seen an automobile. Not only was this the first to enter Valdez, but most of the town's residents had themselves arrived ten years earlier, during the Klondike Gold Rush of 1898, when automobiles had been as scarce in the United States as gold nuggets in the Klondike.

The gold rushers didn't know much about automobiles, but they knew plenty about fool's errands. Certainly enough to guess they were watching one unfold as the *Santa Clara* approached the town pier.

On board, the storm-bedraggled crew of the Thomas stood on deck and contemplated the town and the glaciers and jagged mountains beyond. This was a very different crew than the one that had started out with the automobile back in February. Montague Roberts was gone, having left the race back in Cheyenne, Wyoming, to return east to drive in several upcoming races. Gone, too, was *New York Times* reporter Walter Williams, replaced now by a man named George MacAdam. The only remaining original crewman was George Schuster, the Thomas Motor mechanic who had been called down to New York the night before the race began. With Montague Roberts's departure, Schuster had been promoted from mechanic to chief driver and leader of the expedition. A new man on the team was George Miller, a mechanic for the Thomas Motor Company.

As the attrition of crew members suggested, the journey to this point had been even more difficult than advertised. During the first half of the journey across the continental U.S., both automobile and racers had been frozen by blizzards and entrapped by mud. Since Omaha, they had been sloshed by rivers they had forded, exhausted by mountains they had climbed, baked and choked by the heat and dust of the desert they had crossed, and jostled by the tracks of the Union Pacific Railroad, on which they had driven when no other road was available.

But for all the insults of topography and climate, the longer the race had progressed, the farther ahead of their rivals the Americans had pulled. Eighty miles ahead at Omaha, they were nine hundred miles ahead by the time they reached San Francisco in late March. Even people immune to the charms of automobiles had to admit there was something thrilling about watching Americans best Europeans in this contest. Now the French had lost not one but two automobiles, leaving only the De Dion in the race. The De Dion was in third place, behind the Italian Zust but well ahead of the German Protos, which was somewhere back in Idaho awaiting new parts.

As usual, there was a downside to being first. While the others were back in the States enjoying spring weather, the Thomas crew was at the edge of snowbound Alaska. Nothing they'd undertaken compared to the trail that now lay before them. The plan was that they would drive north 285 miles along the Valdez–Fairbanks Trail, climb over the Alaska Mountain range, then turn west and motor on the ice of the frozen Yukon River to the Bering Strait. It did not auger well that the snow in Valdez was so deep the Thomas Flyer could not be driven into town to take part in the parade arranged to welcome it. The automobile was parked, instead, in a dockside warehouse as its crew greeted the Valdezians on foot.

Around 2 A.M., after another long banquet, the crew of the Thomas set out to survey the trail under the midnight sun, chauffeured in a horse-drawn sled by a local guide. Almost immediately they realized that the trail would not be passable in an automobile. A month earlier, the crust of the path might have supported the weight of the vehicle; now the Thomas would sink to its axles and be stuck until summer. In any case, the forty-five-inch-wide path was much too narrow for an automobile with a fifty-six-inch-wide axle.

Back in Valdez from their sled ride at 10 A.M., the crew immediately made arrangements to return to Seattle. There would be no drive through the Alaskan snows, no drive across the Bering Strait. The race had apparently reached the end of the road.

Big Nail

Even as the American racers were abandoning their ambitions on the far north, Frederick Cook was pursuing his own through arduous conditions. On April 8, the same day the Thomas Flyer arrived in Valdez, Cook and his two Eskimo companions marched into a ferocious and unremitting wind. This was the most difficult weather they'd encountered. Every step and breath became a struggle. "No torment," Cook wrote, "could be worse than that never-ceasing rush of icy air."

The wind stayed with them for days. On the fifth day, one of the Eskimos, Ahwelah, stopped walking and bent over his sled, vowing to die rather than take another step. Cook walked over to him. He saw large tears in the man's eyes. In a "strangely shrilling wail," Ahwelah told him that it was impossible to continue.

"I knew the dreaded time of utter despair had come," wrote Cook later. "I shall never forget that desolate drab scene about us—those endless stretches of gray and dead-white ice, that drab dull sky, the thickening blackness in the west which entered into and made gray and black our souls, that ominous, eerie and dreadful wind . . ." Privately, Cook agreed that the thought of closing his eyes and ending this misery was appealing. He could see how much they all had suffered in the gaunt frames of his companions. He estimated each had lost twenty-five to forty pounds of fat and muscle since leaving Annoatok in February. Their already tawny faces were fissured by wind and sun. "We were curious looking savages," wrote Cook, now including himself, interestingly, among their kind.

What roused Cook's hopes and fueled his exhortations to the Eskimos was his knowledge that they were only one hundred miles from the Pole. Five days. He held up five fingers: five more marches to the Big Nail. "Tigishu-conitu," he assured them. *The pole is near.* He reminded them of all they had accomplished and of how proud their families would be when they returned home. As he spoke he became animated, even vehement. At last, the Eskimos' spirits began to lift. They agreed to push on.

Having recommitted themselves, the three men strode through the remaining miles with renewed energy and purpose. Even the dogs pulled with a new surge of spirit. The hummocks passed in a blur. Cook began to search the distant ice and its vaporous mirages for signs of—what? He did not know. "It now seemed to me that something unusual must happen, that some line must cross our horizon to mark the important areas into which we were passing."

On the morning of April 19, they stopped to rest after another all-night march. They were somewhere beyond the eighty-ninth parallel, less than a degree shy of the Pole. They celebrated with a pot of pea soup, their first hot food in many days. The vista to the north of the eighty-ninth was much like the one they had been looking at for five hundred miles, but it appeared gilded to Cook. They were closing in on the Pole and his imagination was taking flight. "Dull blue and purple expanses were transfigured into plains of gold, in which were lakes of sapphire and rivulets of ruby fire. Engirdling this world were purple mountains with gilded crests."

In these last miles, a "fever heat" of excitement overcame the men's numb exhaustion. The two Eskimos sang as they drove the dogs, who galloped and howled, stirred by the infectious joy of their masters. Despite the fast pace, Cook paused to take sextant readings and to note how the ice now sparkled like diamonds. And then, late on the morning of April 21—"finally, under skies of crystal blue, with flaming clouds of glory, we touch the mark!"

The sextant gave a reading of latitude 90: the geographic top of the world. "We had reached the zenith," Cook later wrote, "of man's Ultima Thule."

> For that moment I was intoxicated. I stood alone, apart from my two Eskimo companions, a shifting waste of purple ice on every side, alone in a dead world—a world of angry winds, eternal cold, and desolate for hundreds of miles in every direction as the planet before man was made.
>
> Over and over again I repeated to myself that I had reached the North Pole, and the thought thrilled through my nerves and veins like the shivering sound of silver bells.

In this heightened state, Cook and the Eskimos turned to housekeeping matters. First came the all-important task of pinning the American flag to a tent pole and planting it in the ice to mark the spot. "I asserted the achievement in the name of the ninety millions of countrymen who swear fealty to that flag," Cook later wrote.

This done, the men prepared camp. As Ahwelah and Etukishook cut blocks of snow for an igloo, they stopped now and then to glance around hopefully. Cook had warned them not to expect an actual iron spike sticking up from the surface of the Big Nail, but they could not help being disappointed that nothing more exotic than ice marked the spot.

For his part, Cook was determined not to let his excitement keep him from taking necessary observations. He gauged the thickness of the ice, measured the shadows, the wind speed, temperature, and barometric pressure. He noted the features of the surrounding icescape, the colors and the "ice-blink" of pearly hues on the horizon. Later, these details would be important in proving that he had indeed gone where he said he had gone. But just where, exactly, was that? The very nature of the Pole made it difficult to say. "That I stood at the time on the very pivotal pin-point of the earth I do not and never did claim; I may have, I may not," Cook later wrote. "In the moving world of ice, of constantly rising mists, with the low-lying sun whose rays are always deflected, such an ascer-tainment of actual position, even with instruments in the best workable condition, is, as all scientists will agree, impossible."

Even as he recorded their position, they were moving away from it. The American flag may have marked the spot at noon, but a few hours later was well beyond it. Unlike the summit of a mountain (for instance), the Pole was a theoretical place, defined by a set of abstract coordinates. Physically, the North Pole—this place he'd just traveled thousands of miles to discover—did not exist.

Cook began to feel a kind of dreadful doubt opening inside of him. Not doubt about the fact that he had reached the Pole (he would never admit to doubting that) but about his faith, so strong only hours earlier, that the achievement meant something. It began to dawn on him—to flood over him like a wave of nausea—that he had just completed a vain and useless act. The real and true accomplishments of humans were not selfish conquests like this one, but in "deeds whereby humanity ben-efits," he later wrote, sounding very much like the progressive he might have been. "Such work as noble bands of women accomplish who go into the slums of great cities, who nurse the sick, who teach the ignorant, who engage in social service humbly, patiently, unexpectant of any reward." The kind of work, in other words, that Jane Addams did every day. Perhaps Cook really had missed his calling.

He had made it to the North Pole, and all he wanted now was to leave. "A wild eagerness to get back to land seized me."

April 21, 1908: "We touch the mark!" Ahwelah and Etukishook at the Pole with the American flag. Photograph by Frederick Cook

*   *   *

As it happened, the very day that Frederick Cook and his Eskimo companions ended their quest at the North Pole, a quest of a very different sort was beginning three thousand miles to the south, in the small Midwestern city of Dayton, Ohio. That evening of April 21, at 6:15, a slight man, unassuming in appearance, boarded a train to embark on a long journey to the Outer Banks of North Carolina. It's unlikely that anyone at the depot recognized him as he bade farewell to his sister and father; only a few would have known his name. That was about to change for all time. When Orville Wright returned to this same depot five months later, under very different circumstances of tragedy and triumph, he and his brother, Wilbur, would be celebrated around the world and Dayton, Ohio, would be famous because of them. Like Frederick Cook, the Wright brothers were heading into uncharted territory. Unlike Cook, they would go there under the eyes—or rather, over the heads—of thousands of spectators who would stand on the ground below, slack-jawed, and wondering at how the world was changing.

PART II

To Conquer Time
and Space

CHAPTER SIX

The Man Bird

APRIL 22–MAY 23

*It was the centre of the world because it was the touchable
embodiment of an Idea, which, presently, is to make the world
something different than it has ever been before. The two little
dots working out there in the sun knew more about this idea
and had carried it farther than anybody else.*

—Arthur Ruhl
Collier's, May 30, 1908

Americans' relationship with the sky changed that spring. In the
past, America had been a nation of people who looked mainly
downward to locate their futures: down at the soil from which they
hoped their crops would grow; down into creek beds for gold nuggets
that might catch their eye and increase their worth; down into mine
shafts for coal and iron ore that would feed their industries. The earth
itself was their greatest source of sustenance and hope. But now, at the
start of the twentieth century came a shift of perspective. The future
beckoned from above, and people flung their heads back to catch it.

A preliminary exercise in mass whiplash occurred on April 22, in
upper Manhattan. This was opening day at the Polo Grounds, the first
home game for the New York Giants in what would turn out to be one
of the most storied seasons in baseball history. Twenty-seven thousand
fans crammed into the horseshoe-shaped stadium to watch the Giants
battle the Brooklyn Superbas. Most sat in the bleachers but many stood
shoulder to shoulder, as per the custom of the day, in the deep outfield,
within spitting distance of the players. Thousands more stood outside

the stadium gates, trying to wedge their way inside, and still thousands
more watched from Coogan's Bluff, the grassy slope that rose above the
stadium, or from apartment windows and rooftops, or from the tracks of
the elevated train—from any perch that afforded even a sliver of a view.

As the game entered the bottom of the ninth, the Giants were down,
2–1. A promising young ballplayer, nineteen-year-old Fred Merkle (the
name would soon go down in baseball infamy), led off with a double.
The Giants took an out on a sacrifice fly, advancing Merkle to third,
then another out when Merkle got tagged trying for home on a double
by Fred Tenney. With two outs and a man on second, the Giants had
one last hope: "Turkey Mike" Donlin. Michael Donlin was one of the
National League's finest hitters (his 1908 batting average would be .334)
but a man of erratic and immoderate temperament, the sort of fellow
liable to pull a gun on a bartender who refused him more drink. (He'd
been sent to jail for doing exactly this a few seasons earlier.) With the
peculiar turkeylike strut that had earned him his nickname, he
approached the plate.

The first pitch came; Donlin swung into a strike. Another pitch;
another strike. Giants fans held their collective breath. "The third, speeding onward, was freighted with the hopes of most of the 27,000 persons
who packed the Polo Grounds enclosure," is how the *New York Herald*
described the next and final pitch. Donlin swung. A telltale crack shot
through the stadium. The sound reached those gathered on Coogan's
Bluff moments after their eyes saw the ball take flight. Twenty-seven
thousand heads jerked back. The ball rose into the afternoon sky on a
splendid trajectory, affirming that baseball was, at moments, fundamen-
tally an aerial sport, in which players and fans remained firmly attached
to earth but all attention and aspiration sailed high above the grass.

By the time the ball landed beyond the right field bleachers, the
crowd was stampeding across the field with "an exhibition of emotional
insanity," according to the *Herald*, "such as is rarely seen on a baseball
field." Donlin just barely managed to dodge his fans as he rounded third
base and escaped into the clubhouse before he could be lifted and carried
away on their shoulders.

A week after Donlin's home run at the Polo Grounds uptown, two
dozen newspaper reporters convened downtown at the corner of Broadway and Liberty Street. They had been invited to tour the new headquar-

Giants fans on Coogan's Bluff, circa 1908

ters of the Singer Sewing Machine Company before the building opened
its doors to the public on May 1. At 612 feet—about 230 feet taller than
any that had come before it—the "Singerhorn" (as wits soon began to call
it) was about to become the highest inhabited building in the world.

Just eighteen years earlier, the highest point in Manhattan had been
the steeple of Trinity Church, two blocks south of where the Singer
Building now rose. Over the last two decades, the most remarkable
building boom in the city's history had transformed Manhattan into a
forest of steel-framed skyscrapers. Older and more timid New Yorkers,
raised in a world of stout masonry structures, still doubted that these tall
wispy skeleton-frame towers could stand a strong wind. Others simply
despised them for their aesthetic intrusion. The writer Henry James,
returning to New York from a long sojourn in Europe—this was a few
years before the Singer went up—found the skyscrapers revolting "mon-
sters of the market" that obliterated from sight his beloved Trinity
steeple. "Where, for the eye, is the felicity of simplified gothic, of noble
pre-eminence, that once made of this highly pleasing edifice the pride of
the town and the feature of Broadway?" wondered James of Trinity's
steeple. "The answer is, obviously, that these charming elements are still

there, just where they ever were, but that they have been mercilessly deprived of their visibility."

James and the worrywarts notwithstanding, most New Yorkers adored the new buildings. Crowds gathered on the sidewalk as the towers rose. They craned their necks to watch the swashbuckling ironworkers— "cowboys of the sky," one journalist called them in 1908—toss red-hot rivets and dance around on the steel, assembling grids of reconstituted iron ore into rooms up high. "Ever and ever upward grows the beetling skyline of the city," remarked the *New York Times* that May. "Will this upward trend eventually make the city a metropolis of cliff dwellers, each looking down into a sunless cañon?"

Scientific American asked how high a building could get, given the limits of physics and New York City's building code. (Two thousand feet seemed to be the approximate answer.) Illustrators like William Robinson Leigh imagined a future city of golden towers connected high above the earth by slender suspension bridges and great masonry arches. Moses King, in a 1908 illustration for *King's Dream of New York,* imagined dirigibles and other flying craft floating over vaulting towers and bridges, bound for faraway destinations like the Panama Canal and the North Pole. "A weird thought of the frenzied heart of the world in later times," King's caption read, "incessantly crowding the possibilities of aerial and inter-terrestrial construction, when the wonders of 1908 . . . will be far outdone."

For the moment, the Singer Building, a Beaux Arts shaft of red brick and blue stone crowned by an elegant mansard roof, was the most dazzling skyscraper that had ever been built. Its height would very soon be surpassed by the Metropolitan Life Building, but not until the Woolworth Building went up in 1913 would the Singer meet its architectural equal. The way it rose so high and slender, then tapered to a cone at the top, called to mind a rocket ship simmering on a launch pad. Of course, no one in 1908 had any idea what a rocket ship looked like, but everyone could appreciate the exhilarating airborne thrust of the building, its missilelike potential. Contemplated from the street below as clouds swam past its cupola, it appeared to be soaring through the sky.

The reporters who had been invited to view the building began their tour by entering a high-speed elevator and shooting up forty-two floors in forty-five seconds. After visiting the observation room high above the city, a few intrepid members of the press corps stepped outside to climb a ladder to the crow's nest at the very summit of the building. Looking west across New Jersey, they could see the Orange Mountains. South offered expansive views over the bay, as far as Sandy Hook. To the north they could see beyond the Polo Grounds into the Bronx and the street-car suburbs of Westchester County. "Those of the party who managed to overcome their positive conviction that the footing must surely give way," wrote one reporter, "had the pleasure of experiencing that thrill peculiar to being poised above nearly seven hundred feet in a gale."

The following evening, a gale, a true one, rushed into the city. Despite the concerns of skyscraperphobics, the steel bracing of the Singer Building held tight in winds of forty-five miles per hour. Lesser structures were not so fortunate. Twenty trees fell in Central Park. Plate-glass windows shattered throughout town. In the East River, a three-masted schooner careened toward the foot of East Twenty-sixth Street and nearly slammed into the riverfront isolation ward of Bellevue Hospital. Uptown, on East Seventy-eighth Street, a woman named Lillian Green was flung by the wind into the path of an oncoming streetcar and only barely escaped with her life.

Whenever gales hit town, the list of casualties always included a few women blown off course. The dead weight of women's corsets and other effects was considerable, but their voluminous skirts filled like sails on blustery days, making them as vulnerable to wind as small craft. Beyond the liability of physical injury, women had their modesty to protect, particularly in the neighborhood of skyscrapers. Tall buildings wreaked havoc on wind patterns and funneled even mild breezes into hem-lifting updrafts.

Women's hats, too, required firm hands in stiff winds. They were enormous and extravagant constructions in the spring of 1908, almost worthy, structurally speaking, of comparison to the Singer Building. "The architecture of the new spring hats worn by women," the *Times* huffed in an editorial that spring, "is amazing." The *Times* did not mean this as a compliment. Festooned with bird feathers, encircled by saucer-shaped brims nearly a yard in diameter, the hats were impractical and exasperating under the best of circumstances. "Properly to display her new headgear the woman of the Spring of 1908 should have an acre of space around her," the *Times* continued. "They are not beautiful because they are appropriate to no occasion unless the place is an open field."

The *Times* was not alone in looking severely upon these ornithological grotesqueries. Audubon Society members correctly pointed out that rare birds had been slaughtered to plume them. Lecture attendees, churchgoers, and sports fans railed against them as impediments to views of podiums, pulpits, and baseball diamonds. Other critics derided them as ridiculous accessories of women who might—who really *should*—be putting their heads to better uses than occupying large hats.

This last point echoed a popular refrain of civic leaders, pundits, and, not least, the president of the United States: that American women at the start of the twentieth century were letting themselves be distracted by frivolities and failing to live up to their God-given virtues and obligations. A woman's first obligation in this view was to get married and bear children. Once ensconced in her role of wife and mother, she was expected to be the conscience of her family, the moral guide to her children, and the exemplar of all that was good and virtuous. Nor did her responsibilities end at the threshold of her home. She was encouraged to apply the same ethics of cleanliness and wholesomeness that she brought to *domestic* life inside to *civic* life outside, thereby achieving "co-operation of good housekeeping and good citizenship," as the editors of *Good Housekeeping* would put it in their November issue. "Women must bestir themselves to purify the city in defense of the home," the magazine advised. "They must combine to make war upon dirt and disease in the street and in the slum in the same way as they now do in the kitchen and the parlor. The town is merely an extension of the home."

Finally, of course, even as she saved her family from moral turpitude and rescued the larger world from grime, and even as she eschewed large hats, she was expected to be beautiful. "A woman," one male authority asserted in *Good Housekeeping,* "has failed to fulfill her first duty to humanity when she fails to be attractive." Translated into the new slang devised by Yale undergraduates that spring (and reported on the front page of the *New York Times*), a young woman was duty-bound to be a "glory," not a "gloom."

Newspapers were filled in 1908, as now, with products and advice to help her achieve her full glory. Along with ads for plastic surgery, beauty potions, and various other cures, the Sunday papers prescribed exercises to reduce unsightly fat. "Standing after meals for twenty minutes, if it will not bring down the hips by pounds, at least aids the digestion, fights off sluggishness and prevents flesh from settling around the waist line," one column in the *New York Times* informed women readers. "Dropping a handkerchief and picking it up without bending the knees, if kept up long enough, is a good reducer for the hips. Unfortunately, it can only be practiced when corsetless, but five minutes morning and evening will do much in the thinning process."

Coincidentally, the same day the *New York Times* reported Yale's new glory and gloom slang (May 6), the newspaper—along with just about every other newspaper in the country—carried the first of many articles about a stout midwestern woman named Belle Gunness. Four bodies had been dug up from the cinders of her torched farmhouse in La Porte, Indiana. Disregarding both her domestic and civic duties, Belle Gunness, the ultimate gloom, had murdered all of her children, then hit the road for better times elsewhere. Mrs. Gunness, it is safe to assume, was not a subscriber to *Good Housekeeping*.

Fortunately, not every American woman chose to follow in Belle Gunness's footsteps and liberate herself by killing her entire family. But given all of the conflicting burdens and responsibilities that befell them, no wonder women might occasionally wish to don enormous aerodynamic hats covered in bird feathers. The hats were like dreams of flight bursting from their heads.

Kill Devil Hill

The first indication that something significant was taking place among the dunes of North Carolina reached the outer world on May 1. A story in a small local newspaper, the *Norfolk Virginia-Pilot*, mentioned that the Wright brothers of Dayton, Ohio, had returned to Kitty Hawk, North Carolina, after an absence of five years, and had flown an airplane ten miles out over the Atlantic Ocean, then back again. The report, like so many regarding the Wrights over the last few years, was nonsense. The truth was that the Wrights had not yet flown ten feet in 1908, much less ten miles.

Orville Wright had arrived at Kill Devil Hill, near Kitty Hawk, on April 25, four days after leaving his father and sister in Dayton. His journey on the Chesapeake and Ohio Railway had taken him along the Kentucky side of the Ohio River—a night ride through Night Rider country—then on through West Virginia and Virginia. In Elizabeth City, North Carolina, he'd picked up supplies that had been shipped ahead, including a disassembled airplane, then journeyed by boat to the Outer Banks of North Carolina.

By the time Orville joined his older brother at Kill Devil Hill, Wilbur had been among the dunes for two weeks, working through a soggy and windy April. With help from a few local men, Wilbur had repaired the old buildings that had fallen into ruin since the brothers' last visit in December of 1903. He'd raised the fallen roof and dug out the old floor from under a foot of sand, then erected a large new cabin where he and Orville could sleep in relative comfort. Charlie Furnas, a mechanic from Dayton, had arrived in mid-April to pitch in. Wilbur, Charlie, and the local men had managed to do most of the work in a matter of days.

On Monday, April 27, the Wrights began unpacking their Flyer from its shipping crate and reconstructing its box-shaped skeleton of spruce and ash. With Charlie's help, the brothers worked side by side. They dressed, as always, in white shirts with starched collars and ties. Both Wilbur, forty-one, and Orville, thirty-seven, were formal men, teetotalers and nonsmokers, careful in their dress and manners though not especially fussy. Their relationship with each other, like so much about them, was fascinating and inscrutable. Neither was married; the only woman in each man's life was their sister, Katharine, with whom they lived, along with their father, in Dayton. Their bachelorhood contributed to their self-sufficiency and allowed them to focus on their work with a kind of priestly devotion. But what alchemy allowed two middle-aged bicycle mechanics from Dayton, Ohio, to achieve a thing at which many men who were wealthier and more educated had failed?

The solution seemed to lie somewhere between Wilbur's more painstaking and lucid intelligence and Orville's more restless and ardent inventiveness. The Wrights were not the unsophisticated rubes they would sometimes be portrayed as in the press. They were both well-read men who grew up in an intellectually brimming household. Both were blessed with a rare combination of mechanical ability, mental rigor, and imagination. They had plenty of differences, of course, and were known to argue vociferously at times, but they seemed always to act as one in the end. Wilbur later insisted that luck played a large role in their success, which would sound like false modesty were it not for the fact that Wilbur seldom said what he did not mean.

One trait both men possessed, and absolutely required, was physical courage. Few innovators risked their lives as consistently as the pioneers of air travel. The history of the development of aircraft had been, as the Wrights knew better than most, a history mainly of failure and death. Each Wright had come close to dying himself a few times. Each would come close again.

By May 4, a week after Orville's arrival, they were ready to test the thirty-five-horsepower engine. "We ran the engine four times, the runs being about 3 min., 5 min., 12 min., and 15 minutes," wrote Wilbur in his journal that day. "It ran each time without stopping."

The entry was typical of Wilbur's unadorned accounts of weather and logistics. Nowhere did he record the emotions he or Orville experienced upon returning to Kill Devil Hill. It's difficult to believe they did not feel any, though, as they looked out over the desolate seascape and smelled the salt air and hint of pine wafting in from trees across the banks. It was to these same dunes that "the boys" (as intimates knew them) had come several years in a row, earlier in the decade, for their "scientific vacations"; where Wilbur had spent long hours sitting on the sand to watch the hawks, eagles, and osprey fly overhead, attempting to glean the secret of flight. Here they had tested their first kites, then climbed the dunes to try their first gliders on swooping rides over the sand. And here, on December 17, 1903, they had achieved the first true powered flight by a heavier-than-air machine, a twelve-second hop by Orville, followed later

Orville Wright   Wilbur Wright

the same day by a fifty-nine-second leap by Wilbur. They had not been back since.

Whatever thoughts stirred inside them now, they betrayed none of them. For men attempting to realize the dream of flight, the Wrights were remarkably grounded. They had little time to indulge in nostalgia in any case. In just a few weeks, they and their machine would begin a season of rigorous and fateful tests. They were here to prepare themselves for what promised to be—what *would* be—the greatest challenge and triumph of their lives.

So much had changed since 1903. The Wrights' ability to fly had advanced almost exponentially beyond those first thrilling seconds in the air. But so had their competition, and so had the pressure upon them to succeed. Back then, the only other American even close to developing a heavier-than-air flying machine was Samuel Langley, the patriarchal head of the Smithsonian Institution. After Langley's "aerodrome" crashed ingloriously into the Potomac River in October of 1903, a couple of months before their own Flyer lifted off at Kill Devil Hill, the Wrights were—momentarily—alone in the air.

At least a dozen eligible rivals now circled them. Most of these rivals were French. Since the eighteenth century, when the French first began to take an avid interest in ballooning, French aviators had considered their primacy in the skies a natural, if not divine, right. This was not all Gallic hubris. Using several airplanes built by a French manufacturer (the Voison brothers), French pilots had begun to fly in earnest. In November of 1906, Alberto Santos-Dumont, a Parisian of Brazilian birth (who had become world-famous after circling the Eiffel Tower in a dirigible in 1901), flew an airplane seven hundred feet. The following year, in November of 1907, Léon Delagrange flew fifteen hundred feet. The latest French success came on January 13, 1908, when Henri Farman, son of an English journalist—but French to the core—won the Deutsche-Archdeacon Prize with a circular flight of nearly two miles, or nearly a minute and a half in the air.

A more recent and ominous challenge to the Wrights came from closer to home. In the fall of 1907, a coterie of several bright and ambitious young men had joined Alexander Graham Bell, inventor of the telephone, to form the Aerial Experiment Association. Bell, now sixty-one years old, had long hoped to solve the riddle of flying. He had devoted his latter years to experimenting with giant tetrahedral kites at his home in Nova Scotia, confident his bizarre creations held the key to human flight. Much to Bell's dismay his acolytes were more interested in building airplanes like the Wrights' than in exploring his geometric contrivances. One of the bright young men, Thomas Selfridge, had even written to the Wrights in January asking advice on aircraft construction, an inquiry to which the Wrights had helpfully replied. Through the winter of 1908, the AEA had designed and built a plane based, at least in part, on Wright ideas. On March 12, 1908, in Hammondsport, New York, Casey Baldwin, another AEA member, had flown the *Red Wing* above an icy lake for a distance of almost 320 feet.

Balloons and dirigibles, too, contended for a place in the aerial future. In fact, at the start of the year, these lighter-than-air craft were the obvious choice for flight. They could remain aloft for hours at a time and travel long distances, and they captured the imagination of Americans. Self-appointed aeronauts crowded the skies that spring of 1908 in various gas-filled contraptions to demonstrate the wonders of floating through the air. Also, too often, they demonstrated the absurdities and dangers. Like boats without rudders, lighter-than-air craft tended to sail along wherever the wind carried them. "Dirigible" was French for steerable, but dirigibles were steerable only in the broadest sense.

Still, to those who had never seen the Wrights fly and had heard only unsubstantiated rumors of their achievements, the flights of the French and of the AEA, as well as of the many balloons and dirigibles, sounded like extraordinary stuff.

The Wrights knew better. They knew, for example, that the accomplishments of their heavier-than-air rivals were based largely on their own designs. This was an irksome but reassuring fact, since it suggested that none had come up with any better ideas. The Wrights also knew that when it came to aerial control—the ability to truly steer a plane—all of their rivals were far behind. The others were taking baby steps while the Wrights were leaping hurdles.

For the Wrights, the object had never been simply to fly. Any balloon could rise into the air if filled with enough hot gas; any fool could strap a strong motor onto a plank and propel it through the air for a few seconds. The real goal was to move through the air with absolute *command*, to turn and soar among layers of altitude as easily and gracefully, and as intentionally, as a bird. The Wrights had come to understand how a complex set of coefficients—propulsion, gravity, air pressure—allowed for the extraordi-narily complex balancing act of flight. When a plane turned, or banked, it simultaneously yawed, rolled, and pitched (i.e., moved laterally, longi-tudinally, and altitudinally). All three of these motions had to work in con-cert if flight, *real* flight, were to occur.

The Wrights' single greatest breakthrough was what they called "wing warping." Slight adjustments to the ends of the wings—bending (or torquing) the tip of one wing *into* the wind while bending the tip of the other wing *away* from the wind—caused lift on one side of the plane and drag on the other. This essential insight allowed the Wrights to turn and roll at the same time, much as a bicyclist negotiates a curve, while main-taining stability. The plane's pitch was controlled by its horizontal nose, which kept it flying smoothly rather than bucking wildly.

The Wrights had won their insights through years of painstaking research. They had built a small wind tunnel to test the variables of air pressure on a body moving through space, and Wilbur had ridden a specially rigged bicycle around Dayton to measure drag and lift on wings. And, of course, the Wrights had built gliders and planes and flown them dozens of times in dangerous but fruitful trials. "There are two ways to ride a fractious horse," Wilbur once wrote. "One is to get on him and learn by actual practice how each motion and trick may be best met; the other is to sit on a fence and watch the beast a while, and then retire to the house and at leisure figure out the best way of overcoming his jumps and kicks." The Wrights had done plenty of the latter, but in the end, wrote Wilbur, "you must mount a machine and become acquainted with its tricks by actual trial."

For two years after their first success at Kitty Hawk, the Wrights had carried out a series of trials at a field outside of Dayton, called Huffman's Prairie. Working as secretively as possible, they had managed to extend their flying times to forty minutes, or twenty-five miles. More important, they'd achieved flights of excellent control, in which they had manipulated their machine through gentle turns and loops and twists, sweeping across the sky as easily as they might have ridden a bicycle through the streets of Dayton. And then, near the end of 1905, they'd abruptly stopped. They had not flown since.

The Wrights had good reason to stop flying when they did. They'd started to attract a lot of attention. And attention, they believed, made them vulnerable to piracy. Not until the spring of 1906 would they have a patent on their technology, and even with that patent in hand they would have to worry that rivals might poach and adapt key features of their design, then claim them as their own.

The brothers came under a good deal of fire for their caginess. Some critics suggested that it was unsporting, at the least, to horde insights that could benefit a great human endeavor like flight. Had the Wrights been saints or millionaires, they might have done as some suggested and donated their intellectual property to the general public. But the Wrights were not wealthy men. They had given everything they had to developing the airplane and had not yet earned a penny. They believed they had every right to profit.

This turned out to be more easily said than done. No government or private enterprise wanted to buy a Wright airplane without first seeing a demonstration of its abilities. But the Wrights refused to demonstrate their plane in public without guarantees of payment up-front. Virtually all aerial exhibitions were public by their very nature. Anyone with a good pair of field glasses could see them.

As the Wrights dallied with interested buyers, first the American government, then the French, skeptics and competitors increasingly construed their reticence as evidence of fraud. The French financier and aviation guru Ernest Archdeacon spoke for many: "If it is true—and I doubt it more and more—that they were the first to fly through the air, they will not have the glory before History," Archdeacon said. "They would only have had to eschew these incomprehensible affectations of mystery, and to carry out their experiments in broad daylight, like Santos-Dumont and Farman, and before official judges, surrounded by thousands of spectators."

In the absence of proof, even the patriotic *Scientific American* worried as late as February of 1908 that Americans lacked the edge in aviation:

> Signs are not wanting that the time is ripe for just such rapid development of the art of navigation in the air, as was witnessed in the development of the automobile, when the French applied their great mechanical genius to that end. And we ask the question: Is the United States to take its proper place as the leading nation in this era of development, or are we to follow along, two or three years behind the rest of the world, and buy our dirigibles and aeroplanes from abroad, just as we were obligated to buy our first automobiles from France and Germany?

The opportunity to prove themselves finally came to the Wrights with two separate offers of purchase. The first of these was tendered, belatedly, by the U.S. Army. It had taken until the end of 1907, but the new Aeronautical Division of the army's Signal Corps had finally grasped the strategic implications of a machine that could be sent high above a battlefield for reconnaissance. The second offer came from a private syndicate which planned to manufacture and sell Wright Flyers in France. Together these deals would not only reward the Wrights financially; they would also serve as a notice to the world that the Wright Flyer was the first true practical airplane in history. There was just one catch: in both cases, the deal hinged on a series of public demonstrations in which the Wrights would be obliged to show that their machine could do everything they claimed it could do.

The Wrights accepted. Wilbur was scheduled to go to France later in the month to begin demonstrations there. Orville would fly for the Signal Corps at the end of the summer near Washington, D.C. The time had come to put up or shut up.

American Genius

Just past noon, on May 6, the Great White Fleet slipped through the Golden Gate and entered San Francisco Bay. The sky was gray, the wind hard and cold, but the largest crowd ever gathered in California's history—half a million people, according to one estimate—stood waiting to greet the ships. Tens of thousands packed the wharves along the waterfront. Tens of thousands more covered Nob Hill and Russian Hill—every elevation around the city—turning them into high pastures of rippling American flags. Under the weight of the crowd and ships, the entire continent, the *New York Herald* declared, "seemed to tip westward."

The *Connecticut* led the way. As the flagship entered San Francisco Bay, guns from the Presidio blasted a salute and torrents of cheers poured down from the hills. The fleet had picked up two additional battleships for the visit, bringing its total to eighteen, along with a flotilla of six destroyers and numerous other accompanying craft. Altogether, the fleet that entered the bay was even more impressive than the one that had departed from Hampton Roads six months earlier. "Hampton Roads is outdone," exclaimed the *Herald.* "The glory of that day is not gone, but it is dimmed in the greater glory of the Golden Gate, three thousand miles away."

Three thousand miles by land, that is, but 14,441 miles by sea. Since the fleet had passed through the Strait of Magellan in February, life for the officers and men aboard had been relatively uneventful. After a quick stop in Callao, Chile, the fleet had anchored in the Mexican port of Magdalena Bay for nearly a month in early spring to take target practice. The ships had been moving at a stately pace up the coast. A few sailors had trashed a restaurant in Santa Barbara, angry they'd been overcharged, but otherwise nothing untoward had occurred. If the voyage had not been as exciting as the press might have wished, it had turned out to be more effective as an advertisement of American naval discipline than President Roosevelt could have dreamed.

The fleet's arrival in San Francisco had special meaning to both the people onshore and the men on board. Not simply because it marked the end of the American leg of the voyage, but because this city, almost exactly two years earlier, had been reduced to rubble by an earthquake. The same hills now teeming with celebrants had teemed in April of 1906 with refugees from the burning city, when all hope seemed to be lost. The rebuilding of San Francisco and the arrival of the Great White Fleet had nothing directly to do with each other, but both were symbols of American will and glory. When San Franciscans gazed over the fleet in the harbor, they saw reflected in its gleaming white hulls their own triumph. "The one example of American genius and energy," observed the *Herald*, "was a fit companion for the other."

About the same time the fleet was steaming into San Francisco Bay that afternoon of May 6, Wilbur and Orville Wright were removing their airplane from its wooden shed and hauling it over the sand to its launch site. The weather was overcast on the Outer Banks, much as it was in San Francisco, but balmier. A mild wind made the day ideal for flying.

A few hundred feet from the shed, the brothers set the machine at the end of a 115-foot steel rail. The wind had been blowing from the southeast, but it shifted a little, so the men pivoted the rail to face directly into the wind. Wilbur slipped into the seat and reached for the levers.

In previous flights, the Wrights had piloted the plane while lying prone, like boys sledding down a hill. Now, partly to comply with Signal Corps specifications and partly to make flying more comfortable on their necks and backs, they'd installed two seats, one for the pilot, the other for a passenger (theoretical at this point, since no passenger had yet flown in a Wright airplane). There were a few other modifications since Huffman's Prairie. The control levers had been changed slightly, and the propellers were now turned by a new, and stronger, thirty-five-horsepower engine.

For men who had not flown at all in three years, these minor changes presented major challenges. The Wrights would have to instantly master the new controls and new power of the plane the moment it lifted from the ground. The margin for error was slim. A single wrong move could injure or kill the pilot, and would almost certainly destroy the plane.

The engine whirred to life. The propellers began to spin on the wings behind Wilbur. For a few moments, the plane remained motionless, held firmly from behind by a cable, tugging forward eagerly like a dog on a chain. Then Wilbur released the cable and the plane started down the track. Orville described the experience of taking off in one of their Flyers in an article he wrote later in the year for *The Century Magazine*:

> Before reaching the end of the track the operator moves the front rudder, and the machine lifts from the rail like a kite supported by the pressure of the air underneath it. The ground under you is at first a perfect blur, but as you rise objects become clearer. At the height of one hundred feet you hardly feel any motion at all, except for the wind which strikes your face. If you did not take the precaution to fasten your hat before starting, you have probably lost it by this time.

Wilbur did not in fact reach an altitude of one hundred feet on this first flight of 1908. He rose about twenty feet and flew a distance of just over a thousand feet. "The flight was very much up and down," he wrote in his journal, "as the operator (W.W.) was thinking more of the side control than of the fore-and-aft control."

The next day, the front page of the *New York Herald* carried a brief story of Wilbur's short flight, squeezed in alongside its abundant coverage of the fleet's arrival in San Francisco. The article was written by a local stringer named D. Bruce Salley. Though Salley did not see the flight, he apparently learned about it from men at the nearby Kill Devil lifesaving station. The article was inaccurate in a few particulars—it suggested that Wilbur and Orville flew together, for instance—but this time, at least, the facts were plausible enough to pique the interest of hardened newspaper and magazine editors around the country. Several of these dispatched reporters to the Outer Banks at once.

From the very start of their May venture, the Wrights understood they'd have limited time to fly before the press caught wind of their presence at Kill Devil Hill. These May days were, in fact, the last moments of their lives when they would live and work in conditions resembling anything like privacy.

On May 10, Byron Newton of the *New York Herald* and William Hoster of the *New York American* arrived in the town of Manteo, on Roanoke Island. At four o'clock the next morning, both men joined D. Bruce Salley for what would become daily excursions to Kill Devil Hill. The trio sailed across Albemarle Sound, then waded to shore and walked several miles through dark swamps and pine woods, to within a mile of the Wrights' camp. There they waited undercover at the edge of the woods, swatting mosquitoes and picking ticks off their bodies. Convinced that the Wrights would refuse to fly if they knew reporters were present, the

Reporters hide in the woods near Kill Devil Hill,
May 1908, waiting for the Wrights to fly.

men were determined to remain hidden and spy on the brothers from afar.

Finally, after a dim sunrise, the Wrights appeared. The reporters watched through field glasses as the brothers positioned the plane on the track and made ready to fly. At 9:16 A.M., the machine, with Wilbur again at the controls, raced down the track and lifted into the air.

This was the first time any of the reporters had witnessed human flight in a heavier-than-air machine. The sight was exhilarating but also professionally challenging. How do you describe something that many have imagined but few have seen? What does a flying machine look like and sound like? How does it work? How does it maneuver through the sky? There were as yet no publicly available photographs, and certainly no moving pictures, and nothing really to compare it to. The reporters did the best they could. "Imagine a noisy reaper flying through the air with a rising and falling motion similar to that of a bird and a fair picture of the Wright brothers' flying machine is obtained," Byron Newton wrote a bit clumsily in the *Herald*.

A few weeks later, he would describe his impressions of this first flight more personally and eloquently:

> The machine rose obliquely in the air. At first it came directly toward us, so that we could not tell how fast it was going, except that it appeared to increase in size as it approached. In the excitement of this first flight, men trained to observe details under all sorts of distractions, forgot their cameras, forgot their watches, forgot everything but this aerial monster chattering over our heads.

Man and Nature

By mid-May, the country was deluged by press coverage of the flights at Kill Devil Hill. But the world did not stop for the Wrights; indeed, other events conspired to draw attention away from them. Harry K. Thaw was back on front pages, dependably weird and petulant. Having decided that insane asylum life was not for him, he was plotting his release from Matteawan. He had managed to obtain a writ of habeas corpus from a Poughkeepsie judge. The insanity plea at his last trial had been "clever," Thaw now told reporters, more of a ruse than a reflection of his mental fitness. He was actually quite sane, he insisted. Certainly he was sane enough to appreciate a good legal strategy when he saw one.

As for that other infamous degenerate of 1908, Theodore Roosevelt, he was up to his usual surprises. His last spring in Washington he was spending the balance of his political capital on a full-blown effort to conserve America's natural environment. It is difficult to imagine another president going to such lengths for such a politically thankless cause. If his tenacity in pursuing such idiosyncratic goals caused some people to think he was crazy, it also commanded admiration.

On the morning of May 13, by Roosevelt's invitation, forty-one of America's forty-six governors gathered in the East Room of the White House for the first-ever Governors' Conference on Conservation. The governors were joined by virtually the entire leadership of the federal government, including all nine Justices of the Supreme Court, most of the president's cabinet, numerous congressmen, and scores of industrial kingpins, labor leaders, and notable scientists—altogether more than 350 prominent men. (And one woman: Sarah S. Platt-Decker, president of the General Federation of Women's Clubs, had been invited to attend.)

Bringing together forty-one governors was no small matter in 1908. Many had traveled days by rail to attend the three-day conference. In return for their exertions, they had been treated to a state dinner the previous evening. Fresh caviar and stuffed squab had preceded cigars and liquors in the White House garden, where Mrs. Roosevelt's flowers bloomed fragrantly, and where the governors were treated to a scene they would not soon forget: President Roosevelt and William Jennings Bryan, the two greatest opposing political forces of the day, deep in friendly conversation amid the roses and clematis. In his every gesture, the president made clear that the matter he'd invited them to discuss transcended partisan disputes. It was a matter of grave national concern, "the weightiest problem," as he put it, "now before the nation."

The events at Kill Devil Hill concerned two men flying through the air in apparent defiance of the laws of gravity. The events in the East Room of the White House involved politicians inflating a room with enough hot air to float a dirigible. Both dramas, though, were variations on a theme that infiltrated virtually every aspect of public thought in early twentieth-century America: the relationship between human beings and the natural world; and the extent to which the former could and should control the latter.

At 11 A.M. the president entered the East Room with Vice President Fairbanks. "You have come hither at my request so that we may join together to consider the question of the conservation and use of the great fundamental sources of wealth of this Nation," Roosevelt began. "We have become great because of the lavish use of our resources and we have just reason to be proud of our growth. But the time has come to inquire seriously what will happen when our forests are gone, when the coal, iron, the oil, and the gas exhausted."

This was not the first time Roosevelt had turned his attention to the environment. During his administration, he had quadrupled the amount of federally protected land to 200 million acres. Indeed, this conference was the culmination of seven years of policy—and the start of a final conservation push before Roosevelt left office and shipped off to hunt wild game in Africa.

The man was forever a paradox. Among the most environmentally friendly presidents in history, he was also the most bloodthirsty.

The president's desires were not as contradictory as they might seem. Roosevelt considered nature a crucible at which a man might sharpen his instincts and build his muscles and discipline. Wilderness was the playground of natural selection. Like his old friend John Muir, founder of the Sierra Club—an organization devoted to preserving the environment in an unadulterated state—Roosevelt worshiped nature. But where Muir worshiped in silent and passive regard, Roosevelt worshiped by slaughtering animals and laying their bloody corpses on the altar. Roosevelt was not a preservationist, he was a conservationist; to him, the goal was to conquer nature yet not destroy it. Because, then, what would be left to conquer?

Roosevelt's practical concerns were different than those of preservationists, and different, too, than those of later environmentalists. He and his fellow conservationists didn't worry that the earth was being fouled by pollution; they worried it was being stripped by *depletion*. Coal reserves and minerals, forests and fresh water, would be used up if America did not take care, they warned. "Our geologists give but a few centuries before our coal supply will have become exhausted," one conservationist wrote in the *New York Times* later in the year. "The exhaustion of our iron and oil will be forerunners of this great calamity, and then mankind will be brought face to face with that terrible situation—savagery."

Many years would pass before Americans began to worry as much about the waste that came *out* of America as the waste that went *in*. But if the rationale of an early twentieth-century conservationist was different than that of an early twenty-first-century environmentalist, the imperative was the same: to use less. And to Roosevelt's everlasting credit—again, it's hard to imagine many politicians doing this—he was addressing concerns not of the politically urgent present, but of a future he would not live to see.

"We must handle the water, the wood, the grasses, so that we will hand them on to our children and our children's children in better and not worse shape than we got them," he told the governors. "Any right thinking father earnestly desires and strives to leave his son both an untarnished name and reasonable equipment for the struggle of life. So this Nation as a whole should earnestly desire and strive to leave to the next generation that national honor unstained and the national resources unexhausted."

When the president finished, the governors responded with enthusiastic applause. The room was filled with friends and foes, but all joined, at least for a few moments, in adulation of this extraordinary force of nature who was soon to be their ex-president.

The governors join President Roosevelt for the conservation conference, commenced May 13.

*   *   *

Kill Devil Hill was alive with activity that same morning of May 13. Shortly after dawn, the Wrights had appeared with their plane. The pool of reporters lying in wait had now grown to half a dozen men, including P. H. McGowan of the London *Daily Mail*, and writer Arthur Ruhl and photographer Jimmy Hare, both of *Collier's* magazine. Ruhl would later publish a humorous description of the journalists' predawn trek to Kill Devil Hill, comparing their attempt to spy on the Wrights to a military reconnaissance mission. "The shortest way, of course, would have been to climb up one side and down the other, and thus descend directly on the beach and the aeroplane camp," wrote Ruhl. "And then there would have been no flights that day. We must needs, therefore, act exactly as if a platoon of sharpshooters were in trenched on the other side, with their fire raking the summit of the slope, turn to the left and make a wide detour to gain the timber on the farther side."

The reporters watched Wilbur step into the plane and take off. Ruhl's colleague, Jimmy Hare, jumped up to snap the first-ever publicly distributed photographs of a Wright plane in flight. ("Don't shoot till you see the whites of their eyes," shouted McGowan, a veteran war correspondent.) Later in the morning, Orville took the controls of the plane and flew the longest flight yet at Kill Devil Hill, nearly three minutes in duration. This was another minor outing compared to the flights the Wrights had made at Huffman's Prairie in 1904 and 1905, but to the newspapermen it was plenty impressive. Byron Newton had his wits about him this time. The story he filed for the next day's *Herald* is worth quoting at length, not only because it is a stirring account of a remarkable day, but also because it put the world on notice that something fundamental had changed in humans' relationship with the sky.

> With the ease and swiftness of a huge eagle the Wright brothers' aeroplane made a flight of three miles at ten o'clock this morning, circling about the great sand hills, at times skimming along over the surf, dipping down, rising, turning corners, and landing within a few yards of the starting point. . . .
>
> As it sailed along to-day, flashing in and out among the glittering sand hills, one must have imagined it to be a huge bird in soaring flight, except for the constant roar and rattle of the motor, and somewhat rigid directness of its course. . . .

Had the performance taken place in some great arena with excited crowds cheering all around, probably the charm and the magic of the thing might not have appealed to one's senses so strongly, but here on this silent beach, with the blue ocean for a background, there seemed to be something supernatural about this mysterious aerial traveler as it soared along, its incessant chatter mingling with the sullen booming of the surf. At all times the operator could be plainly seen, bending and turning as he manipulated the levers. With a twist of one lever one great wing of the machine would tilt up, and away it would shoot to the left or the right. The twist of another lever, and it would dart downward or upward, the movement being more easily accomplished than the manipulation of an automobile under moderate speed.

Newton's words, appearing in the *Herald* and picked up by countless newspapers around the world, carried a great deal of weight. "There is no longer any grounds for questioning the performance of the men and their wonderful machine," wrote Newton. "In short, their flight to-day will practically change the whole aspect of things in the field of aerial navigation."

*   *   *

Taken at Kill Devil Hill on the morning of May 13 by Jimmy Hare of *Collier's Weekly*, this is probably the first publicly distributed photograph of a Wright airplane in flight. (The plane is the small dot near the center of the photograph.)

The following day, May 14, was clear and sunny. At 8 A.M., reporters were treated to the first-ever two-man flight aboard a Wright plane. Wilbur took Charlie Furnas on a short hop. Then Orville took Furnas up for a flight of almost four minutes.

That afternoon, reporters observed a more astonishing spectacle. Wilbur was up alone and had been flying for almost eight minutes—the longest flight of the spring. He had turned one large circle at an elevation of about thirty feet and was entering into another circle. Just as Byron Newton was judging the flight "the most important achievement in the history of aerial navigation," the plane suddenly vanished behind a dune. The reporters heard its engine go suddenly silent.

Orville was watching through field glasses when he saw the plane drop, nose first, about a mile from camp. An instant later, he saw "a big splash of sand—such a cloud as that I couldn't see from where I was exactly what had happened." To which he added, with typical Wright understatement, "We were somewhat excited at camp for a few seconds."

Wilbur did not appear from the wreckage for almost half a minute, but when Orville and Charlie Furnas finally reached him they found him to be in fine shape. His face was scratched, his hands and body were bruised, but he was otherwise intact. The same could not be said of the plane. It lay on the sand in shambles, beyond repair.

The flight trials at Kill Devil Hill, after only a few days of actual flying—fewer than two dozen short flights—were over. The next time either man flew, he would have to wing it, truly, in front of the world.

On May 17, Wilbur left Kill Devil Hill, en route to France via New York City. Orville remained behind to finish packing up their camp. That same day, Henri Farman, the now-famous French aviator, publicly bet the Wrights five thousand dollars that he could beat them in both speed and distance. Coming directly on the heels of Wilbur's crash, Farman's challenge was a pointed reminder that the Wrights remained vulnerable to competition and that their rivals were ready to pounce.

The point was reiterated the following day, May 18. In Hammondsport, New York, the AEA resumed its own flying trials in a new and more sophisticated airplane, the *White Wing*. With the addition of so-called ailerons—bendable sections at the tips of the wings that resembled the warped wings of the Wrights—Casey Baldwin flew ninety-three yards in a straight line. On May 19, Glenn Hammond Curtiss, also of the AEA, flew the *White Wing* 340 yards, with several turns and considerable control.

All of this may have been discomfiting to the Wrights, but to the rest of the world it was thrilling. The airplane was not a fluke ginned up by two quiet bachelors from Ohio. It was a widespread innovation that was meeting sudden success throughout the world. As recently as April of 1908, skepticism had been the best policy regarding heavier-than-air technology; by the middle of May, flight was an indisputable reality.

"Man has flown," wrote George Grantham Bain in *Hampton's*. "Conqueror of the earth and water, he has made the air a medium of motion—he has sustained himself on his own man-made wings and, rising from the earth, he has circled through the almost intangible ether." Bain understood that what the Wrights had accomplished was only a start. "The man bird of to-day will seem as awkward and futile to the man bird of a hundred years hence as the first type of steamboat appears to the builder of the *Mauretania*."

On the foggy morning of May 21, Wilbur boarded the *Touraine* for a seven-day crossing to Le Havre, France. Wilbur probably did not see the hull of the *Lusitania*, sister ship of the *Mauretania*, as he embarked from New York, but the *Touraine* passed the famous Cunard liner in the fog later that morning. The *Lusitania* was just completing a record-breaking voyage from Europe, having crossed the Atlantic on the "long route" (from Queenstown, Ireland) in four days, twenty hours, and twenty-two minutes. No ship, not even the *Mauretania*, had ever gone so far so fast. Surely not even Wilbur imagined that someday man-birds and woman-birds would cross the Atlantic in a matter of hours.

A Bird! A Bird!

The first strong bit of advice is that every girl should realize that the street is a public thoroughfare, and that everything she says and does is noticed. . . . She should remember that a lady should never draw the slightest attention to herself in public. She must moderate her tones, be quiet in her actions.
—*The New York Times*, May 24, 1908

If two men flying through the air in a motorized airplane were not enough to convince Americans that they were living in unusual times, the fact was confirmed a few days later, on May 23, when a pretty young chorus girl named Bertha Carlisle stepped out onto State Street in downtown Chicago wearing a dress that nearly started a riot. The dress Miss Carlisle wore was known to the French as the *directoire*. Americans called it a "sheath." What mattered about the dress was not its name, but its tight-fitting, hip-hugging form. Its most explicit element—explicit being a term relative to 1908—was a slit that ran from the floor-length hem to the knee. The slit was hardly revealing since the dress was worn with leggings underneath, but the consensus, as voiced by the *Chicago Tribune*, was that it was "daring, sensational, disquieting." It was possibly the most sexually charged garment that had ever appeared on a public street in an American city.

The gown had first entered public discourse earlier in the month, on May 10, when three young French women appeared wearing directoires at Longchamp racetrack in Paris. That appearance had caused such a sensation that French police removed the young women from Longchamp to prevent a riot or an accident. Four days later, on May 14, a young equestrienne galloped through the Hyde Park section of London dressed in a sheath and very nearly did cause an accident. A young man named Winston Churchill happened to be trotting by in the opposite direction. Turning in his saddle to watch the woman pass, he violently collided with another rider. On the same day that Wilbur Wright

Two of the sheath-clad French women who scandalized Longchamp on May 10, 1908. Two weeks later, Bertha Carlisle wore a similar dress in Chicago.

barely escaped death in his airplane, then, the future Prime Minister of Britain nearly got killed by a dress.

And now it was America's turn to meet the perilous gown. Miss Carlisle had purchased a directoire while on a visit to Paris. The manager of her theater company, a man named Joe Weber, had bet her that she would not dare to wear it in public. She declared she would, and the press was duly notified.

"There never was anything in feminine apparel more comfortable," Miss Carlisle assured reporters before she set off on her stroll down State Street. "The absence of petticoats and furbelows, perfect freedom of the limbs, and clinging material, make it an ideal dress for the woman possessing a figure." The thousands of men who mobbed her as she stepped onto State Street attested to the fact that Miss Carlisle possessed a figure.

She began at 4:30 P.M. from the Stratford Hotel, accompanied by a business manager of the Colonial Theater, as well as an entourage of plainclothes police officers and reporters. "The clinging fabric was all curves, and fitted Miss Carlisle's figure exactly," the *New York Herald* reported. "Every time one of Miss Carlisle's ankles was displayed from among the folds of her gown some observer coined a new term to voice his approval."

Flashes of blue satin leggings, glimpsed through the slit of the dress, sent men into paroxysms of delight and women into fits of contempt. "My," said a fat woman, "I wouldn't wear a gown like that for a hundred dollars." In fact, Miss Carlisle was wearing it for *five* hundred dollars, the amount she had paid for it in Paris—and the size of her wager with Joe Weber.

Miss Carlisle stopped in at a few stores, at first to preserve the illusion that there was something natural about her stroll down State Street, then simply to catch her breath away from the growing melee. "The male rubbernecks came up on a run from all points of the compass," reported the *Chicago Tribune*. "Women fought madly to get a view of the creation. Gloves were torn. A woman's skirt was ripped up her back. Another lost her belt, another her hat. Coiffures were hacked to pieces—all except those of Miss Carlisle, whose nodding gray plumes on top of her thick black hair were visible in the crush. . . ."

Realizing too late the whole thing might have been a mistake, Miss Carlisle sought refuge from the growing riot inside Marshall Field's. As

the crowd attempted to follow her into the department store, she slipped
into an elevator and repaired to the tranquility of the ninth floor. She
applied powder to her face and fixed her hatpins. Then she went back
out to face the crowd.

"A bird, a bird!" cried one man when he saw her. "That's what she is!"

"O, it was a dreadful experience. I felt as if I'd faint when all those
men kept crowding round me," Miss Carlisle told reporters after she'd
collected herself and her five hundred dollars. "I was terribly frightened.
I wished I was under a bed or could wrap my head in a United States
flag."

CHAPTER SEVEN

Heat

JUNE 15–JULY 24

Our hearts go out with Peary . . . and many wish that our bod-
ies could go with him too.

—*The New York Times*, July 24, 1908

Three weeks after Bertha Carlisle's sheath-clad walk down State Street, all of Chicago bloomed with United States flags. It was June 15, Flag Day by proclamation of the governor, and nine hundred delegates were converging on the city for the Republican National Convention. Buildings throughout the city flew Stars and Stripes to greet the delegates. At the Chicago Coliseum, a huge stone edifice on Wabash Avenue built to resemble a medieval castle—all the better to ward off berserking Democrats—flags luffed on every turret. Inside the enormous hall, more flags hung from the gray stone walls and tricolored bunting swaddled the steel rafters.

The convention to elect the next Republican candidate for president was really over before it began, its conclusion preordained by the man who had been nominated in this same hall four years earlier. President Theodore Roosevelt had handpicked William Howard Taft, his secretary of war, to succeed him, and there was very little that the other six candidates could do about it. Not even Taft, who did not especially want to be president, seemed able to resist Roosevelt's will.

In the days leading up to the convention, there remained just one possible spoiler to Taft's lock on the nomination: Theodore Roosevelt himself. Here was an aspect of the convention beyond the president's control. He had sworn up and down that he did not wish to serve a third term; had categorically refused to serve a third term; and had insisted that any talk of a third term was preposterous and insulting to his character. But rumors of a Roosevelt "stampede" refused to die.

The closer Roosevelt came to the end of his presidency, the more his fellow Americans—some of them, anyway—meant to hold on to him. So thoroughly had he remade the office in his image that the idea of the office continuing without him was, to many, a doleful and bewildering prospect. Even those who were not particular fans of the Roosevelt presidency were bothered by his impending absence. "The old house will seem dull and sad when my Theodore is gone," wrote Henry Adams, an old acquaintance and frequent critic of the president.

Already the old house seemed dull and sad at times that June. Most of the president's six children, having grown into teenagers and young adults over the family's seven years in the White House, were gone, scattered on independent summer ventures. Fortunately, to Roosevelt's delight, his youngest son, ten-year-old Quentin, remained at home to keep the place from becoming too quiet. Quentin was a sprightly, charismatic boy, and his father's favorite. As Roosevelt worked in his office, and as the most beautiful Washington spring in years thickened into a hot Washington summer, Quentin played baseball or mock war games on the White House lawn with his usual gang of local friends. The president occasionally marched out to intervene, ostensibly when the games became destructive of government property, but more likely to join in the childish fun for a few moments. ("You escape, Quentin," he scolded his son after Quentin cut a White House hose with an ax, "only because of extenuating circumstances arising out of the heat of battle.") At night, after Quentin went to bed and the sounds of laughter and chatter died down, the rooms came to feel, Roosevelt wrote, "big and lonely and full of echoes."

It was not easy for Roosevelt to relinquish the White House. Seldom has a man enjoyed an occupation or occupied an office so completely as Theodore Roosevelt did the presidency. Having technically served only one full term (he'd entered office in September 1901 after President McKinley's assassination), he could have tried for another, and very likely would have won it. But Roosevelt knew better than to give into the temptation to stay. "While President I have *been* President emphatically," he wrote in a letter in June. "I believe in power; but I believe that responsibility should go with power, and that it is not well that the strong executive should be a perpetual executive."

So now he was in the peculiar position of working to ensure his own eviction from a home and job he loved. To ease the sting, he'd found in Taft "a man whose theory of public and private duty is my own, and whose practice of this theory is what I hope mine is." Roosevelt believed sincerely—and naïvely, it turned out—that Taft, who looked a bit like an oversized version of Roosevelt (who already looked like an oversized version of a lesser man), would happily carry on as his political doppelgänger. "[I]f we can elect him President we achieve all that could be achieved by continuing me in the office, and yet we avoid all the objections, all the risk of creating a bad precedent."

The convention opened on Tuesday, June 16. Unlike previous years, no portraits of Republican fathers adorned the walls. The most conspicuous portrait to appear during the opening ceremonies was on a large banner unfurled by the delegation from Taft's home state of Ohio. The portrait was meant to show a likeness of Taft but was so poorly rendered that many people thought it was Roosevelt. Which may have been exactly the point.

The following afternoon, June 17, as sun streamed through the glass transoms in the vaulted roof of the Coliseum, Roosevelt's old friend

As the Republican convention begins in Chicago, the June 13 issue of *Harper's Weekly* shows Taft trying to squeeze his famous girth into Roosevelt's famous Rough Rider outfit.

Senator Henry Cabot Lodge of Massachusetts took the floor as convention chairman. Lodge had come to Chicago determined to silence any
talk of a Roosevelt stampede, but shortly into his speech the Senator
himself brought down the house by referring to Roosevelt as "the best
abused and most popular man in the United States today." The words
acted, according to the *New York Tribune,* "as burning fuse to powder."
At first, as the *Chicago Tribune* reported, the applause "rose and wavered,
and almost died away," but then a voice cried out, "Three cheers for
Teddy!" and the volcano that had been rumbling and simmering suddenly exploded. What came then was a "deafening, throat-splitting,
heart pulsating exultation," wrote the reporter from the *New York Herald.*

As Lodge attempted to bring the hall back to order by holding up his
hands, the noise grew in pitch and volume. "Four more years! Four more
years!" chanted the crowd. From his White House office, President Roosevelt listened to the outburst through a telephone that had been hooked
up to receivers in the Coliseum. If the ovation was tactically problematic,
sounding a lot like the start of the stampede Roosevelt had sought to
quash, the outpouring of affection must also have been profoundly gratifying. Roosevelt had suffered a barrage of political attacks ever since the
crash of 1907, most of these launched by members of his own party. Today,
came only praise. "President Roosevelt's most consistent enemies cannot
deny the spontaneity of the applause," reported the *New York Herald.*

For about fifteen minutes, the delegates on the floor of the convention
hall joined in the shouting and stomping, standing on chairs and waving
hats and canes. Gradually, the delegates began to quiet, ready to carry on
with the business at hand. This had little effect on the volume in the
room. Most of the noise came from the galleries above the floor, where
about eleven thousand spectators stood in bleachers, hollering themselves breathless. The crowd in the galleries had no vote and no authority at the convention. What it had was voice, and it refused to quit. "Had
the galleries had their way," wrote the *Herald* reporter, "they would have
nominated Roosevelt unanimously."

Half an hour into the ovation, Senator Lodge pounded the gavel. His
efforts brought only more noise. "If it was a hurricane before," reported
the *New York Tribune,* "now it was a cyclone." From somewhere a giant
teddy bear, the size of a ten-year-old boy, appeared in the hall, tumbling
from delegation to delegation. The band began to play in hopes that
blaring brass might drown out the crowd, but the crowd simply drowned

out the brass. Alice Longworth, the president's newly married daughter, sat onstage, beaming with pride.

At last, after forty-seven minutes, Lodge managed to bring enough order to the hall to speak. Anyone who encouraged a third term, he remonstrated, was "no friend to President Roosevelt." The crowd gradually quieted. There would be no stampede. The greatest ovation that had ever been heard in American political history was over.

The following day, Taft was nominated by an overwhelming majority of delegates, 702 of 976, on the first ballot. The ovation for Taft lasted a mere twenty-nine minutes, but the deed was done. Whatever happened now, Roosevelt would not be serving a third term. "There is a little hole in my stomach when I think of leaving the White House," wrote ten-year-old Quentin. For the rest of the country, Roosevelt's departure would create a very big hole that Taft's girth could not possibly fill.

On June 20, President Roosevelt left to spend the summer at his home in Oyster Bay, New York. He would not be back at the White House until fall, when the general election would decide whether his successor was to be William Howard Taft or William Jennings Bryan. By the following spring, he and his family would be gone; and those early summer days Roosevelt passed in the White House listening to Quentin play war games on the lawn would be fading into bittersweet memories.

President Roosevelt and his wife, Edith, as they appeared in the *New York Times* on June 14, 1908.

The Lure of the North

The Roosevelts arrived in Oyster Bay just as the first true heat of summer struck the Atlantic seaboard. Temperatures in Washington, D.C., soared on June 21. In New York City, leaden humidity added to ninety degrees of mercury to bring the hottest day since the previous summer—"a heat which struck down man and beast beneath its blast," the *New York Times* reported.

For the wealthy, who could decamp to second homes in cooler environs, summer heat waves were unpleasant but bearable. The prosperous middle class, too, had escape options. They could vacation at one of the sprawling hotels in the Green Mountains of Vermont or on the shore of Long Island—a place like the Manhasset House on Shelter Island, for instance. Hundreds of airy rooms, a private golf course, long distance telephone service in every chamber, and (as advertised) an impressive "thirty miles of splendid roads for automobiling and driving" were just a few of the hotel's amenities.

Automobiles provided another measure of purchased relief to those who could afford them. In those days before air-conditioning, fast driving was the next best thing to wind. "In hot summer months," *Harper's Weekly* rhapsodized to its affluent readers, "there is a delicious refreshment in driving after dark, when the rush of the car through the air creates a cool breeze, and the roadway no longer reflects the blinding rays of the sun."

The working class and poor, of course, were in no position to enjoy either the mountain air of Vermont or the delicious refreshment of an automobile ride. Trolley cars and shaded parks were more likely reprieves. So were the city's rivers. Boys stripped and jumped into them, swimming in their coolness. There was risk in this. Not only were the rivers filthy, but their currents were devilish, and boys often drowned on summer days.

Smaller children lined up in the broiling sun at the doors of the Association for Improving Conditions for the Poor, which provided free baths. Four at a time, the children were dipped in water, cleaned, patted dry, then returned to the hot streets.

Evenings, after the sun went down, were hardest to endure. To escape crowded and stifling tenements, families slept on roofs or fire escapes, or they dozed in parks until a policeman nudged them with a nightstick and sent them along. Some people rode the streetcars all night just to feel the breeze through the open windows.

Finally, after several days of severe heat, the weather cooled slightly in New York. June 28 brought a patchy, capricious day. Downtown, the sky was clear, but uptown rain squalls opened up over Harlem. One side of the Bronx was fair all day, the other drenched. Adding yet another level of meteorological whimsy, the moon partially eclipsed the sun that morning, casting a strange pall over even the cloudless precincts of the city. Street vendors sold pieces of smoked glass to those who wished to view the eclipse. Boys on the Lower East Side saved their pennies by viewing it in murky puddles of old rainwater.

Two days later, in an unrelated event, about six thousand miles from New York, over a remote area of north-central Siberia, something came dislodged from the firmament and fell to earth.

The Thomas Flyer and its crew of Americans happened to be driving through Siberia in late June. They had arrived in the Russian vastness weeks earlier, by way of Japan and Manchuria.

After their futile attempt to penetrate Alaska in April, the Americans had hurried back to Seattle, then crossed the Pacific Ocean by steamship. By the time they reached Japan, the race they had entered was no longer the race they were driving. The elimination of Alaska and the Bering Strait from the route had deprived the contest of its most stirring components. But rather than terminate the race, officials went back to the map and redrew it. The abridged plan called for the racers to travel to Japan, then ferry across the Bay of Japan to Vladivostok. From here, they would continue west by automobile through Manchuria and Siberia.

The Thomas and its crew had arrived in the port of Yokohama, near Tokyo, in the middle of May. Given tensions between the U.S. and Japan, the Americans had landed with some trepidation. This was allayed soon enough—the Japanese, with one or two exceptions, were unfailingly polite—but the drive across the island nation was harrowing anyway. The ancient cart paths, narrow and winding, were not designed for automobiles, and automobiles were not designed for the steep hills. The Thomas had neither the horsepower to ascend the escarpments of the interior nor the brakes to safely descend. The Americans had no choice but to hire local Japanese men to help them tow the car up and down the mountain paths with ropes.

From the western shore of Japan, the Americans boarded another boat and crossed to Vladivostok. There they found their fellow contestants idling in various states of disrepair and disillusion. The crew of the De Dion, the only French car remaining in the race, had just gotten word that its sponsor was withdrawing the automobile and selling it, having apparently concluded that the race was a waste of resources. The Italian manufacturers' support for the Zust crew was likewise waning. Practically speaking, this left the Americans and the Germans. The Italians would eventually continue, but not until weeks later.

A race that had lost half of its original contestants and scrapped much of its original itinerary was arguably no longer a race worth winning. To make matters worse, the race's organizers had awarded the American crew a fifteen-day credit (for its futile excursion to Alaska) and docked the German crew a fifteen-day penalty (for transporting its automobile across much of the northwestern United States by freight train). This thirty-day adjustment made sense as a matter of fairness. As a matter of sport, it leached whatever suspense or logic remained in the race.

The only people who evidently believed they were in a meaningful contest by this point were the contestants: the three men in the German car, the four men in the American car. Ironically, as interest waned elsewhere, the race became truly a race for the first time—a neck-and-neck struggle between the Americans and Germans in the far reaches of the earth. They carried on as if the thirty-day gap did not exist; as if the fate of the world depended on the outcome of their driving.

After departing from Vladivostok on May 22, the Americans and the Germans leapfrogged through the muddy bogs of Manchuria, then took to the tracks of the Trans-Siberian railroad. Driving with one set of tires on the inside of the tracks, the other set on the outside, they thudded along hundreds of bone-jarring miles over the raised wooden ties, their discomfort made more acute by fear of the fierce brigands known to roam Manchuria and by the prospect, always real, of a freight train barreling toward them around the next bend.

Entering Siberia, they returned to roads which were uniformly terrible but no worse, the drivers agreed, than most American roads. They roared through corridors of pine and cedar and larch, the great Siberian taiga. The distances were mind-boggling, the mosquitoes intolerable. During those few hours of night when they paused for rest, they napped fitfully in barns or filthy inns. They subsisted on a diet of eggs and black bread.

On the morning of June 30, the Thomas was on the road to Omsk, three days ahead of the Protos. At 7:15 A.M., in central Siberia, a thousand miles to the northeast of the Americans—a stone's throw in Siberian terms—a bluish ball of light flew across the sky. Moments later, according to local natives who lived to tell of it, a concussion rocked the air for hundreds of miles around, knocking people to the ground.

What occurred that morning in Siberia remains a mystery to this day. The Tunguska Event, as it has come to be known (because it occurred near the Tunguska River) was apparently caused by a large astronomic mass entering the earth's atmosphere. Most likely, modern scientists speculate, a giant meteorite exploded miles above the earth, sending an "airburst" down upon the forest and leveling it with the force of a large nuclear bomb. Whatever caused the event, the result was a blast zone encompassing 830 square miles and as many as 80 million felled trees. For days afterward, skies above Europe glowed eerily.

The heat was back in New York by dawn of July 4. Anyone who could afford to leave the city had already beaten a hasty exit, not just from the scorching temperatures but also from the impending Independence Day revelries. To middle- and upper-class New Yorkers, the Fourth of July was a menacing extravaganza of violence and drunkenness. Firecrackers exploded, bonfires burned, revolvers shot out through day and night. It was as if the poor, left to their own devices, laid siege on the city. The Society for the Suppression of Unnecessary Noise printed the national statistics: 1,153 killed, another 22,520 seriously injured over the last five Independence Days. "What an indictment of our common sense!" railed Rhode Island's *Providence Journal,* among other newspapers around the country. "Why do we allow this slaughter and maiming to go on? Why is it that in the enlightened twentieth century we are so wedded as a people to a barbarous celebration of the Republic's holiday?"

Whether due to a sudden contagion of common sense or to the thundershowers that threatened most of the day, July 4, 1908, turned out to be relatively tranquil in New York. Casualties were limited to 253 injured—a hundred fewer than in 1907—and five dead. Another five corpses showed up in the rivers by the end of the day, including that of a seven-year-old boy in the East River. These could not be fairly attributed to the nation's birthday, though, as boys and young men often died in the rivers on hot days.

Among the more peaceful and safe diversions of the Fourth was a visit to a pier at the far end of East Twenty-fourth Street. All day perspiring throngs visited Robert Peary's ship, the *Roosevelt*. The ship was docked on the East River under the scorching sun, as Peary and his nineteen-man crew made final preparations for their imminent expedition to the North Pole.

Peary's departure came two days later, on Monday, July 6. The heat was stupendous. Before eight, the temperature had reached eighty-two degrees Fahrenheit with a humidity level of 78 percent. It rose steadily from there. The *Herald*'s thermometer soared to ninety-eight. According to the *Tribune*, the heat "made life almost unbearable, drove several persons insane, prostrated nearly a hundred and killed half a score."

As thousands gathered again on the pier on East Twenty-fourth Street to see them off, the crew of the *Roosevelt* worked through the morning heat to load on supplies. Journalists and visitors stepped aboard to poke around the quarters of the ship. They inspected the two-hundred-volume library in Peary's cabin, considered the large provisions of flour (sixteen thousand pounds) and corned beef (twenty

Robert Peary aboard the
Roosevelt on his previous
try for the North Pole.

thousand pounds), and pitied the shaggy sled dogs tied up under the port rail, panting in the heat.

At noon, a young woman drove up in an automobile to deliver several more packages of books to the *Roosevelt*. A little girl then presented the captain with two volumes of fairy tales. Among the last loads to be taken aboard the *Roosevelt* were large cakes of ice. Many were struck by the irony of a ship bound for the North Pole taking on ice.

Peary arrived shortly after noon. Accompanied by his wife and five-year-old son, he obliged the crowd with a few final words before departing. "I am not foolish enough to say that I am going to do or die," he told them, "but I am certainly going to put into this trip every bit of energy, mental, moral and physical, that I have." A day earlier, interviewed in his hotel room by a reporter from the *New York American,* he had spoken more bluntly of his ambitions: "We must win this time."

Robert Peary was over six feet tall, with a large barrel-chested frame and steely blue eyes. At fifty-two, he still exuded robust vigor, though in fact he was well past his physical prime. His earlier expeditions had not only robbed him of his toes, they had also wrecked his health. He had given the greater part of his adult life to pursuing a kind of frozen chimera and had suffered for it terribly. Now he was off to pursue it once again.

Peary traced his ambition to reach the North Pole to childhood. Like so many boys of his generation, he'd read Elisha Kent Kane's memoir of his 1853 journey to the far north and had been haunted by it. Of course, the vast majority of American boys who read Kane's book about the far north had not devoted their lives to going there. Robert Peary, like Frederick Cook, possessed that quirk of nature that makes a person need to do such a thing. In Peary's case, the determination to achieve his end was epic. Anyone familiar with Herman Melville's *Moby-Dick* might have recognized in Peary a drive worthy of Captain Ahab. If he was not as mad as Ahab, he was every bit as obsessed. Whereas Ahab needed to plunge a harpoon into the great white whale, Peary could be satisfied only by sinking Old Glory into the great white of the north. He frequently touted patriotism as his guiding motive in reaching the Pole—"the thing which should be done for the honor and credit of this country"—but mainly it was the thing which had to be done for Peary. "Remember, mother," he had written home as a young man, "I must have fame."

He had in fact achieved considerable fame by 1908. What he lacked was demonstrable success. Since first visiting Greenland in 1886, he had led six more expeditions to the Arctic, two of these failed attempts on the Pole. On his most recent try, in 1906, he had gotten farther north than any man, a mere 174 miles from the Pole, before his supplies ran out and he was forced to turn back. It was, he wrote, "the greatest disappointment of my life." He started planning his return on the boat ride back home. "The lure of the North!" he wrote. "It is a strange and powerful thing."

Arranging that return had taken more effort than he'd anticipated. For all his acclaim, he found money hard to raise. After his longtime benefactor, the New York banker Morris K. Jesup, died in January of 1908, he was forced to go begging where he could. No one wanted to fund another unsuccessful effort. Always the question arose—could he do it this time? Peary was sure that he could.

He intended to approach the Pole on a route familiar to him from earlier expeditions, and the most logical by which to mount a "determined, aggressive attack." Following the so-called American Route, the *Roosevelt* would steam north from the Labrador Sea, entering the Davis Strait, Baffin Bay, and finally Smith Sound, between Ellesmere Island and northern Greenland. In Greenland, he would find the dogs and Eskimos he needed to go further. After passing the long polar night at the edge of the Arctic Ocean, he would make his final approach from the northern tip of Ellesmere Island.

Peary had been offered the use of numerous contrivances to help him get to the Pole. One man volunteered a submarine by which Peary might reach his destination under the polar ice. As to how Peary might drill *through* the ice when he got there, the submariner remained vague. "There was an incredibly large number of persons who were simply oozing with inventions and schemes, the adoption of which would absolutely insure the discovery of the Pole," Peary wrote. "Naturally, in view of the contemporaneous drift of inventive thought, flying machines occupied a high place on the list." Peary dismissed all of them. "Airships, motor cars, trained polar bears, etc., are all premature except as means of attracting public attention," he wrote. Peary believed that with good sledges, good dogs, and handpicked Eskimos, his way offered the best chance of reaching the Pole. And he was sure that his experience gave him a better chance than anyone else.

What he could not be sure of, in early July of 1908, was how close his former first mate and current nemesis, Frederick Cook, had already gotten to the Pole. He knew Cook was trying for it—the news had been in the papers the previous fall—but no word had come from Cook in months. Cook's friends had begun to worry that he was dead. At the very least, Peary had to hope and believe, he had failed.

It was one o'clock in the afternoon when the *Roosevelt* backed out from the pier at East Twenty-fourth Street, drawn by a tugboat. Rather than turn south for New York Harbor, the tugboat pivoted north, in the direction of Long Island Sound, with the *Roosevelt* still in tow. Vessels that had gathered on the river tooted farewell. Sugar factories and powerhouses blasted their whistles from shore. As the *Roosevelt* passed under the great new cantilevered bridge to Queens, inmates incarcerated on Blackwell's Island (now Roosevelt Island) waved from the riverbanks.

On board the *Roosevelt*, iced drinks and luncheon were served to the hundred or so guests who had been invited to join the crew for the first few miles of the voyage. At City Island, the tug halted and unleashed the *Roosevelt*. The guests boarded the tug for the ride back to Manhattan. Peary, too, stepped off the *Roosevelt*. Rather than spend the night in his cabin on board, he returned to the city with his wife and son aboard a yacht owned by the *New York Herald*.

The following morning, Peary and family entrained to Oyster Bay, on Long Island, to lunch with the president. The day was even hotter than the previous one, but the president and Peary wore formal whites. The president, always stirred by adventure and exploration (he would make his own voyage of discovery, to the Amazon, in 1913) did not have to feign interest in Peary's expedition. Peary had been savvy enough to name his boat after the president as an additional guarantee of presidential favor. Roosevelt reciprocated by lending his infectious enthusiasm to the project, and later by writing the introduction to Peary's memoir.

After lunch, the president and his family accompanied Peary to the *Roosevelt*, now docked at Oyster Bay. For an hour, as the sun baked down, the president poked around the boat, shaking every hand and declaring everything "bully!" Before returning to Sagamore Hill, the President stood with the explorer on deck to pose for a photograph. "I believe in you, Peary," said Roosevelt, "and I believe in your success—if it is within the possibility of man."

President Roosevelt bids
Peary farewell on the deck
of the *Roosevelt* at Oyster Bay
on July 7, 1908. The president's
son Quentin stands to the
left of Peary.

"Mr. President, I shall put into this effort everything there is in me—physical, mental, moral."

"Well, Peary, good-bye, and may you have the best of luck."

Dreamland

Even as he bid farewell to Peary, another voyage must have weighed on the president's mind that afternoon. Several hours after the *Roosevelt* steamed out of Oyster Bay, the Great White Fleet passed through the Golden Gate without ceremony and sailed from San Francisco Bay. The first leg of the fleet's voyage had taken it through friendly waters; now it was headed across the Pacific, by way of Hawaii, to Japan and unknown circumstances. Tensions between the U.S. and Japan had eased since the winter, but they were by no means forgotten. That very Sunday, the *New York Times* ran a feature entitled "If War Should Come with Japan," in which the author imagined a grim scenario involving "our Pacific Coast . . . overrun by the armies of Japan, our Pacific squadron sunk, our vaunted battleship fleet surprised and defeated by the ironclads of Togo." A popular novel published earlier in the year, *The Vanishing Fleet* by Roy Norton, helped to fuel the paranoia with a fictional account of just such an attack.

From Denver, meanwhile, came news that William Jennings Bryan had been chosen as the Democratic nominee on the first ballot, an outcome so predictable it barely registered as news. On this, his third and final try for the White House, the "Great Commoner" had abandoned his defining issue—switching America from a gold standard to a silver standard currency—and assumed a platform which sounded remarkably like the policies of Roosevelt. "Will the American people have the Roosevelt policies continued and administered by Mr. Taft or by Mr. Bryan?" wondered the *New York Times.* "That is the real question before them for decision." The *Times,* for one, preferred Taft. Roosevelt as channeled by Bryan, a fundamentalist demagogue in the view of the *Times,* would make for "a veritable reign of terror."

In New York City, the heat continued to press down with an insistence that made all thoughts of November elections impossible. Day after day, the mercury climbed. Horses toppled in the streets, where they remained until the ASPCA came along and hauled their bloated corpses away. The air reeked of putrefying flesh and human sweat. A fine dust of dried dung and ash wafted through open windows and settled over every surface. There were no fans to move the air in the tenements of the poor, no indoor plumbing from which to draw a cool bath. Newspapers tallied the human casualties. Twelve dead one day, twenty the next; scores more prostrations. Adding a note of menace to the infernal heat were reports of attacks by rabid dogs. Rabies, or "hydrophobia," as doctors called it, was an especially horrible death in the heat of summer. It cursed its victims with quenchless thirst, then sent them into convulsions when they tried to drink. After several people died of rabies, the health department ordered all stray dogs shot on sight.

On the night of July 16, at about 11:30 P.M., millions of tiny white moths spontaneously appeared in Harlem and began to flutter downtown in a cloud. "They were so thick under street lights the city appeared to be under a snowstorm," reported the *Times,* conjuring a wishful mirage in the heat. Entomologists were consulted but could provide no immediate explanation for the invasion. Reporters speculated that the moths had come from across the Hudson, floating over from the hills of New Jersey. Wherever they originated, they were in Times Square by midnight, seeking the lights inside the Times Building and bothering the compositors trying to put together the morning's paper on the sixteenth floor. The moths continued south as far as City Hall, then vanished as quickly as they had appeared.

Half a million New Yorkers descended upon Coney Island that weekend. On the night of Saturday, July 18, twenty-five thousand people slept on the beach. Many, desperate for relief from the heat, remained bobbing in the water all through the night.

The heat was unendurable. And yet, for those who had no choice but to endure it, the city offered compensatory pleasures. The poor lacked Bailey's Beach at Newport, but they had Brighton Beach in Brooklyn. If they could not sail yachts on Long Island Sound or sniff the mountain air of Vermont or drive automobiles in the evening hush, they could at least take the trolley to Coney Island and ride Shoot-the-Chutes at Luna Park or the Mountain Torrent at Dreamland. They did not have polo matches, but they had Polo Grounds, where they could sit in the bleachers for twenty-five cents and watch the baseball season unfold.

A splendid season it was, too. The New York Giants were shaping up as pennant favorites, and New Yorkers flocked to upper Manhattan on the subway or on the Sixth Avenue elevated. Some of the fans, no doubt, arrived whistling the melody of a new song by a Tin Pan Alley lyricist named Jack Norworth and composer Albert Von Tilzer. Norworth had never been to a baseball game in his life, but riding into the city on a train one day in 1908 he saw a billboard that read BASEBALL TODAY—POLO GROUNDS. Inspired, he immediately scratched down some lyrics about a baseball fan named Katie Casey with a taste for peanuts and Cracker Jacks. The insanely catchy tune, promoted by "pluggers" who sang it to audiences at nickelodeons and vaudeville houses, was an instant hit. "Take Me Out to the Ball Game" would go on to become one of the most familiar songs of the twentieth century.

Nickelodeons were another pleasure enjoyed by the poor and working class in 1908 but still generally avoided by the middle class and wealthy. Since 1905, as many as eight hundred theaters had registered with the Department of Buildings in New York City, several hundred of these in Manhattan alone. "The popularity of the moving pictures has increased so much in the last year," reported the *New York Tribune* in 1908, "that from one to two places are now on every street on the East side." Some were large, legitimate theaters. Most were ad hoc structures: a projector, a piece of canvas stretched across a wall, a door in back. The newspapers warned that the theaters were fire traps. The movie projectors had a regrettable tendency to explode; the heavy decor a tendency to ignite. Still, New Yorkers continued that summer to pile into the dark theaters to enjoy flickering lights and shadows. "You see what it means to them," one contemporary wrote of moviegoers. "It means Opportunity—a chance to glimpse the beautiful and strange things in the world that you haven't in your life [and that] would be forever closed to you." The danger only made the experience more thrilling, particularly the kind of danger that progressives warned against. When Jane Addams wrote that "the very darkness of the room" encouraged the "glamor of love making," she meant this as a caution against movies, but no better advertisement could have been written to promote them.

A peculiar irony attended the lives of the proletariat. At work, they suffered almost medieval conditions of hardship and bondage. But in their leisure time—Sundays and the brief hours after work—they led America into the twentieth century, tasting certain of its fruits (with the notable exception of the automobile) long before their wealthier compatriots. The rich still clung to nineteenth-century values and imported European culture; the poor were already embarked on an exhilarating ride into modern America. What did the future hold? A visit to a rich man's mansion on Fifth Avenue, dressed up like an English manor house or a French château, would tell you little. Better to find your way to the Lower East Side, or to Harlem or the Tenderloin. The future was prophesied in the syncopated music wafting out of dance halls and taverns—precursor to jazz—and in the frank raw "tough dances," like the turkey trot and the bunny hug, practiced inside.

As the poor embraced new entertainments, keen entrepreneurs were quick to grasp the potential of a new indigenous consumer culture. Individually, the poor had only nickels to spare, but together they were worth untold millions. Fortunes could be made from their nickels, assuming a way could be found to appeal to their modern appetites. As consumers, their every whim and wish would be catered to, as if they were not paupers, but kings.

Coney Island was their kingdom.

In the mythology of American immigration, the first icon encountered by the huddled masses as they arrived on American shores was the

Statue of Liberty. In fact, Coney Island dazzled the immigrants long before Lady Liberty swung into view. The three great amusement parks on the southern shore of Brooklyn—Steeplechase, Luna Park, and Dreamland—defined a vision of the new world more evocatively than any statue could have done, no matter how grand or well-meaning. All the more evocative should the immigrants' ship happen to arrive on a clear night, when the parks were illuminated by millions of lights and their domes and minarets glimmered thirty miles out to sea.

A daylight visit to Coney Island could not possibly live up to that sublime nighttime vision, but it came close. For the price of a trolley car ride and an admission ticket, a visitor could sample the delights and adventures of the world. A gondola ride on a Venetian canal, a journey over the Alps, a tour of Paris—Coney Island offered simulacrums of all these. A visitor could watch a naval battle between America and Japan, take a submarine excursion to the North Pole (like the one Robert Peary had turned down), or go aloft on a make-believe journey in an airship. Those who wished to dance could do so in the largest ballroom in the world, lit by ten thousand bulbs and more dazzling than the ballroom of the Waldorf-Astoria.

Freak shows were another Coney Island specialty. Puerile and demeaning, they also served a psychologically meaningful purpose. The midgets of

Luna Park at night in 1908

Midget City, or the giants—or the sexually ambiguous, the grotesquely fat and ugly, the folk who dwelled on Freak Street—all challenged the metrics of everyday life. This was the very essence of Coney Island. It removed its patrons from the expectations of humdrum reality and threw them into a strange new world of mirages and vertigo. Space and time collapsed as completely as they did at the North Pole. Day and night were mulled into a kind of boreal twilight. Coney Island stayed open twenty-four hours, and visitors who arrived in the early hours of a Sunday walked past bleary-eyed revelers returning home from a late Saturday. The only governing principle was desire—for adventure, for speed and fear, for food and flesh. Over the course of a day, a visitor's blood would be flushed from head to foot—and to loins, too, for the place was charged with sex. Many of the newer rides, like the Human Roulette Wheel or the Human Niagara, were expressly designed to toss people into concupiscent heaps. At Steeplechase's Insanitorium, jets of compressed air shot up from the floor and lifted women's skirts. The Tunnel of Love carried couples into dark subterranean places, their bodies scented by ocean salt and human sweat.

The distinctly modern experience of Coney Island was made possible by cutting-edge technology. The rides were marvels of engineering—the best of twentieth-century science harnessed in the interest of bricklayers and tailors, sweatshop girls and domestics. The Witching Waves, a mechanism of such complexity that *Scientific American* dispatched a correspondent to investigate it, was a fine example. This new Luna Park attraction had been designed by Mr. Theophilus Van Kanel, famed inventor of the revolving door. To provide riders the stomach-churning sensation of being tossed about on stormy seas, Mr. Van Kanel employed small synchronized electric motors to heave scalloped metal plates in undulating patterns.

The impresarios who ran Coney Island were determined to keep their patrons constantly enthralled with new simulations and stimulations. If one of the entertainments failed to please the people, it would be gone next year, replaced by another that would, management most solicitously hoped, impress and delight them to their satisfaction. The customer was king and hero: pioneer of the future, conqueror of the air, discoverer of the North Pole.

"Our hearts go out to Peary, and many wish that our bodies could go with him too," mused the *New York Times* on July 24, as the city continued to swelter. "He is steaming northward to face danger and privation, but he has escaped something it was worth his while to get away from."

Three weeks out of Oyster Bay, the *Roosevelt* was nearly at the Arctic Circle. The journey was off to a good start. Unlike three years earlier, when Peary and his crew had been hampered by thick fog at this point in the voyage, the air remained clear. The ship entered the Davis Strait, heading toward Baffin Bay, and the days grew longer and the air colder.

The *Times* urged its readers who were stuck in the sweltering city to indulge in an imaginative exercise. "Let us try to think about Peary, the whitened coasts of Greenland, the frozen sea beyond, the attractive icebergs," the newspaper suggested in an editorial. "It is folly to wish ourselves with him. We cannot all discover the North Pole. But we may find some comfort in thinking about him in his present environment."

CHAPTER EIGHT

The Color Line

JULY 15–AUGUST 24

Abe Lincoln brought them to Springfield and we will run them out.
—Resident of Springfield, Illinois, August 15, 1908

Somewhere west of the Ural Mountains, on the edge of the eastern plain of Europe, the Thomas Flyer began to falter. The vehicle's collapse occurred in sorrowful stages. First to go was its transmission. Having been subjected to countless topographical and meteorological insults—snowdrifts, mud slicks, potholes, railroad ties, mountain inclines, swamps, deserts, and forests—the transmission's gears were worn to nubs. The American crew had taken heroic measures to preserve them, in one instance smearing the entire apparatus with forty pounds of Vaseline (no other lubricant being available at the time), but all to no avail.

The end came one afternoon in mid-July after the Thomas got stuck in a boggy pothole near the town of Viatka. George Schuster applied the gas, the Thomas's engine whined plaintively, but the gears failed to engage; they were stripped.

The loss of the transmission was a hard blow to the crew of the Thomas. Until this moment, the Americans had managed to dominate the race to Paris, consistently leading the U.S. leg of it, then holding a convincing, if sporadic, lead on the Russian leg. Once they crossed the Urals, the hardest part was supposed to be behind them. The land was flat, the roads were improved, the accommodations and people more refined. As it happened, though, Europe was the Americans' undoing.

Before they could undertake repairs on the automobile, George Schuster had to travel to another town to acquire a new transmission, then haul the six-hundred-pound part back to Viatka by horse wagon. On his return journey, Schuster took a moment's rest in a roadside cabin. He was sipping a cup of tea when he heard an automobile approach. Given the scarcity of automobiles in Viatka, he guessed at once who it was. He did not even bother to look up as the Germans roared by.

The Americans lost five days in Viatka. After installing the new transmission and resuming the race, they compounded their delay by getting themselves frequently and disastrously lost on the outskirts of Moscow. They blamed their navigational problems on the Russians' inability to understand hand signals, but their own sleep-deprived faculties probably contributed to the communication breakdown. They were driving day and night, eating little, resting less, and their brains were half scrambled by the constant rattling of the chassis.

The Thomas arrived in Moscow at 4 A.M. on Tuesday, July 21. The Americans might have continued straight through the city were the automobile not in urgent need of more repairs. As soon as these were complete, they lit out for St. Petersburg, night-driving through corridors of dark forest. When they roared into towns, they found the streets alive with Russians who had stayed awake to cheer them. Onward through the night and the following day, and the following night, they sped after the Germans, trying to close a three-day gap. News came that the Protos was near Berlin; it was still possible to catch up. Barely pausing for breath in St. Petersburg, the Americans made another all-night dash, skirting the southern shore of the Baltic Sea and entering Germany on July 25. The radiator immediately sprang a leak—another sixteen hours lost to repairs, and the final, fateful blow. The following day, when the Germans entered Paris, the Americans were still east of Berlin.

Once it became clear to the Americans that they would not pull ahead of the Protos, they slowed their pace and attempted to enjoy the remaining miles of Europe. The countryside was beautiful, the roads superb, the roadside crowds friendly. The American crew might regret their second-place finish on the leg from Vladivostok to Paris, but at least they were guaranteed victory in the longer race. Between the fifteen-day credit awarded them and the fifteen-day penalty charged to the Germans, they could stroll into Paris twenty-nine days after the Protos and still win.

The Thomas stumbled once more outside of the German city of

Hannover. This time it was a coupling shaft that broke. For twelve hours, George Schuster and machinist George Miller repaired the automobile under a roadside apple tree. Finally, on the evening of Thursday, July 30, the Thomas entered the suburbs of Paris from the northeast, speeding over the cobblestone roads at sixty miles per hour. In the town of Meaux the Americans were met by a convoy of French automobilists who had driven out from the city to greet them. The Thomas fell in line behind its escorts for the last twenty-five miles, running through streets filled with cheering crowds who flushed out of cafés and homes to see them.

The Americans arrived in the center of Paris around 8 P.M. to find the streets filled with still more well-wishers. Boulevardiers who had been enjoying the summer evening packed around the automobile to inspect and applaud the four very dirty and exhausted men who sat inside. The Thomas's slow progress through the crowd was halted near the Place de l'Opéra by a gendarme. "You are under arrest," he called to the Americans inside. "You have no lights on your car." Somewhere on the miles of hard road, the headlight had been destroyed or lost, and now the Thomas had entered Paris in flagrant defiance of a city ordinance. The crowd appealed to the gendarme to consider extenuating circumstances—the car had just driven across Siberia! It had come all the way from New York!—but all of this was immaterial to the gendarme. Finally, a generous bicyclist came to the Americans' rescue. He offered his headlamp to light their way. When it proved difficult to remove the lamp, the bicyclist slung the whole bicycle into the front seat of the Thomas and jumped into the back, and off they went on their victory lap of Paris.

It was victory, but a mystifying, unsatisfying kind of victory. The Americans had arrived at the finish line in *second* place at the end of a race that was not the race they had set out to run. The route, the automobile, and the crew (with the exception of George Schuster) had all been overhauled extensively. Like the philosophers' riddle of the wooden ship, in which every plank of a ship's body is replaced one by one, the race begged the question: at what point did it cease to be itself and become another race?

The *New York Times*, for one, would not allow ontological inquiries to muddy the contestants' achievement. "All agreed in declaring the performance the most remarkable ever undertaken in the history of sport," declared the *Times*. The paper went on to assert that the race should be

taken as evidence of American prowess in automotive manufacturing—
"a great triumph for the American industry."

More thoughtful was a *Times* editorial a few days later.

> The self-moving car is to play a very prominent part in our future history.
> It is to be used for freight as well as passenger traffic, to supplant the
> trucks and vans drawn by horses, to be applied to farm machinery, to be of
> service in war, if, as we hope, the wars of the world have not all been
> fought. The way in which the Thomas and the Protos have accomplished
> the seemingly impossible task set for them indicates that the hope of
> inventors and the commercial and engineering world in the future of the
> automobile has not been misplaced.

If the automobile was only to be "a mere toy" used sometimes "as a
quick and spectacular means of self-destruction," its value would be limited. What the race made clear, the *Times* concluded, was that the age of
practical automobile travel had arrived.

The *Times*'s estimation of the future importance of automobiles was
prescient. Its assumption that the race somehow advanced this, however,
was dubious. The automobiles had been unimpressive as a group. Half of
the original six had dropped out, and the Thomas, the best of the lot,
had averaged just 107 miles a day. The machines were fickle and balky,
and the infrastructure to support them—roads, gas stations, garages—
was screamingly inadequate. Judging by the success of the automobiles,
an impartial observer might have reasonably concluded that the practi-
cality of the automobiles for long-distance travel was decades away.

The truth is that the race was the untidy end of a chapter rather than
the start of a new one in automotive history. It was, for all the talk of heroics and science, little more than a well-publicized lark. The future would
deliver something quite different; and the future was nearer than anyone,
in that midsummer of 1908, might have guessed. In the meantime, the
Americans' victory, legitimate or not, was a symbol to be savored among
the many signs that the nation was becoming the technological leader of
the world—on the seas, in the air, and on the roads. And though the victory had nothing to do with this, it happened to come at the very
moment America was poised to take control of the exploding automobile
market.

The most lasting contribution of the race was not practical or even symbolic, but mythological. Even as the machines proliferated over the next years and decades, evolving from playthings of the rich into afford-able conveyances of everyman, automobiles would always be something more than means of transportation. They would be vessels of imagination. Just as the road movie genre would become a staple of Hollywood (and would include a 1965 farce based on the New York–to-Paris race, *The Great Race*), the cross-country pilgrimage, with its numerous pit stops, breakdowns, and adventures along the way, would become an American rite of passage. No machine ever had or ever would serve so well the pecu-liarly American impulse to hit the road and go forever, to recapitulate the ways of our pioneer ancestors by speeding into the horizon. The New York–to-Paris race helped ensure that automobiles would always be something more than mere machines.

The crew of the Thomas Flyer remained in Paris for a week after reaching the finish line. Then George Schuster and George Miller drove out along the Seine to the port of Le Havre, where they boarded the French steamer *Lorraine*—the same ship that had brought the European cars to New York way back in February. On August 8, with the Thomas Flyer secure in the ship's hold, the Americans embarked for home.

Il Vole

The same day the Thomas Flyer and its crew left France, Wilbur Wright flew for the first time since arriving in the country two months earlier. Wilbur had endured a difficult summer. He had come to a foreign coun-try to demonstrate a machine to a skeptical public that did not especially want him to succeed. The May flights at Kitty Hawk may have satisfied Americans regarding the Wrights' achievement, but to the French they were provocations. French aviators responded in a snarky combative spirit. First, Henri Farman had challenged the Wrights with his five-thousand-dollar wager. Then, on May 30, the day after Wilbur arrived in France, Léon Delagrange had flown nearly ten miles—more than fifteen minutes aloft—at Rome, setting a record for the longest publicly wit-nessed flight. Wilbur feigned indifference when reporters asked him about Delagrange's record. "We are not worried," he told them. But this was not entirely true.

Wilbur had spent his first days in France scouring the countryside near Paris for an appropriate place to stage his demonstrations. He set-

tled on the town of Le Mans, 125 miles southwest of Paris, where the
Hunaudières racetrack would make a suitable, if not ideal, flying ground.
An old acquaintance, the automobile manufacturer Léon Bollée, offered
Wilbur use of his factory near Le Mans to assemble the plane as he pre-
pared to fly. Wilbur moved into the factory and got to work.

To outside eyes, Wilbur presented, as always, the picture of cool self-
possession. The reality was more complicated. A series of peevish letters
he wrote to Orville that June brimmed with anxiety. At first, he
reproached his younger brother for failing to write him and give news of
his progress in America. "I am a little surprised that I have no letter from
you yet," he wrote on June 3, only four days after arriving in France. A
week later, having still received no letter from Orville, he wrote to their
sister, Katharine. "Does he not intend to be partner anymore? It is ridiculous to leave me without information of his doings and intentions."

In the middle of June, Wilbur opened the crates in which Orville and
their assistant, Charlie Furnas, had packed the disassembled plane
Wilbur was to fly. His mood went from bad to worse. Inside the crates,
he found a tangle of cracked spars and torn fabric. Some pieces of essential hardware were missing; others were loose at the bottom of the crate.
"I opened the boxes yesterday and have been puzzled ever since to know
how you could have wasted two whole days packing them," he wrote to
Orville with barely restrained fury. "I am sure that with a scoop shovel I
could have put things in within two or three minutes and made as good
a job of it. I never saw such evidences of idiocy in my life."

Whoever deserved the blame for the poor packing job (Orville laid it
on careless French customs officials), the result was several weeks of
unanticipated work for Wilbur. He had to rebuild the Flyer almost from
scratch, working ten hours a day in Léon Bollée's automobile factory.
The time Wilbur lost was not merely irksome; it was potentially ruinous.
After years of hiding their light under a bushel, the Wrights were in a
sudden rush to exhibit it to the world. For all the good publicity that had
come out of the May flights, the brothers were in danger of forfeiting
their primacy in aviation if they did not soon demonstrate, conclusively
and publicly, all they had achieved. It was necessary, as Wilbur wrote to
Orville, "to let the general public become accustomed to linking these
ideas with us before others attempt to steal them."

Every week, a new achievement by an aeronaut not named Wright
was reported in the press. The French were flying their airplanes with

great success. The Germans, led by Count von Zeppelin, were sailing a giant new dirigible, the LZ4, on impressively long flights over the German countryside. And in America, on July 4, Glenn Hammond Curtiss flew the AEA's third and finest craft yet, the *June Bug*, a kilometer and a half over a field in Hammondsport, New York. That Curtiss's *June Bug* might have infringed on Wright patents mattered more to the Wrights than it did to anyone else.

That same July 4, injury was added to insult. Wilbur was working on his plane's motor when a faulty radiator hose burst and sprayed his torso and arm with scalding water. The burn was exquisitely painful and potentially dangerous, forcing Wilbur to postpone, once again, his demonstrations. The French press was scornful. *Le bluffeur*, they implied, had found another opportunity to delay. Orville was only slightly more sympathetic, reminding Wilbur that Henri Farman had just arrived in America to demonstrate *his* plane at Brighton Beach. "If you don't hurry up," Orville wrote to Wilbur, "he will do his flying here before you get started in France."

On August 5, Count Zeppelin's LZ4 caught fire and self-immolated in its gases. To observers like the editors of *Scientific American*, the accident confirmed the inherent instability of lighter-than-air craft. The future of flight was with the airplane. But whose airplane would it be?

Late on the evening of August 6, under cover of darkness so as to avoid the eyes of lurking journalists, Wilbur and his assistants towed the Flyer by automobile from Léon Bollée's factory to the racetrack at Hunaudières, where they parked it in a shed. Wilbur would live in this shed for the next month, sleeping beside his plane, taking his meals at a local restaurant and his milk from a nearby farmhouse.

Neither the plane nor Wilbur's burn was entirely repaired by August 8, but a rumor had circulated that Wright intended to fly that day. By afternoon, a small but significant crowd was gathered at the track, including two Russian officers dispatched by their government to scout the plane. Much of the French aviation establishment was in attendance as well. The pilots Léon Delagrange and Louis Blériot (who would be the first man to fly across the English Channel in 1909) were present, as was the financier—and Wright archnemesis—Ernest Archdeacon. All had come to watch Wright fly. Or better yet, to not fly and fall on his face.

"I thought it would be a good thing," Wilbur later wrote to Orville, "to do a little something."

*   *   *

Wilbur spent the day inside the shed preparing the airplane. Finally, in late afternoon, he and his assistants rolled it out onto the racetrack. Many minutes were spent checking and testing equipment and patching a small tear in a wing. For the spectators, this was the first taste of the man who would perplex and fascinate them for months to come. He was fanatical in his preparations. People who came to watch Wilbur fly would usually find themselves waiting for hours as he strolled around the plane again and again, whistling softly, testing cables, trying the controls, running the engine and listening to it with detached concentration.

Now, under Wilbur's direction, the plane was set astride the usual steel rail for its launch. To help the plane achieve the necessary speed for takeoff, its nose was attached by a taut cable to heavy iron weights that dangled from the top of a thirty-foot derrick. When the weights were dropped—the effect was rather like a guillotine—the plane would lurch forward.

It was six-thirty in the evening when Wilbur finally climbed into the seat and took hold of the wooden control levers. He wore his usual gray suit, starched white collar, and green cap turned back so it would not blow off in flight. The evening was calm, and so, outwardly, was he. But his heart must have fluttered just a little. This would be the very first public demonstration of a Wright plane. Much, possibly everything, was riding on it. A mistake would be fatal to the brothers' reputation, if not to Wilbur himself. The last time he had flown, he had crashed and destroyed the plane. If he did so now, the French trials would be over before they had begun, and the name *Veelbur Reet*, as they pronounced it in Le Mans, would be the punch line of a French joke.

Spectators watched from the grandstand as the twin propellers behind Wilbur started to spin. All at once, the weights fell from atop the derrick and the plane shot forward on its track. Four seconds later, it was airborne, rising quickly to an altitude of thirty feet, higher than most of the French had flown but low enough to give the audience a view of Wilbur as he made a slight adjustment to the control levers. The plane instantly responded, one wing dipping, the other lifting. The plane banked to the left in a tight, smooth half-circle. Coming out of the turn, the plane made a straight run down the length of the racetrack, about eight hundred meters, then banked and turned into another half-circle. Wilbur looped the field once more in a long oval, then brought the plane down almost exactly where he had taken off less than two minutes earlier.

It was a brief flight, a mere clearing of the throat compared to others Wilbur had made. But those hundred or so seconds were arguably the most important either Wright had spent in the air since their first flights of 1903. With those hundred seconds, everything changed.

Wilbur could see the effect as soon as he landed. Spectators were running across the field to shake his hand, including the same French aviators who had dismissed him only recently as a charlatan. They were almost swooning now as they raced each other to congratulate him. More than anyone, they understood that they had witnessed a level of aerial competence far beyond anything they had achieved. Blériot was speechless. Delagrange was beside himself. "Magnificent! Magnificent!" he cried out. "We're beaten! We don't exist!" Another French aviator declared, "We are as children compared to the Wrights." Even Ernest Archdeacon, who had never had a kind word to say about the Wrights, expressed remorse for doubting them.

The French press was equally magnanimous in the next day's papers. "It is not without a certain sadness that we applaud Mr. Wright's success," confessed an editorial in *L'Intransigeant*. "The aeroplanes of Farman and Delagrange would seem to be only copies. When we contested the sincerity of Mr. Wright we were trying to believe that France would lead in aviation . . . but Mr. Wright proves that we are but debutantes."

"The Wright brothers now should be as much felicitated as they have been scoffed at," declared *La Presse*, while another French journalist wrote, "That man Wright has conquered the air." At the very least, he had conquered France.

Wilbur never flew on Sundays. But on Monday, August 10, he took off again. This time, a crowd of thousands was on hand. He aborted his first attempt almost immediately, but on the second he flew two figure eights, an unprecedented maneuver in aviation, and he did it to perfection. The response was even more overwhelming than it had been two days earlier. "Bleriot & Delagrange were so excited they could barely speak," Wilbur wrote Orville. "You would have almost died of laughter if you could have seen them."

To make these August days sweeter for the Wrights, things were not

Wilbur at Hunaudières, August 1908.
"I thought it would be a good thing
to do a little something."

going well for that other French aviator, Henri Farman, the man who
had bet the brothers five thousand dollars he could outfly them. His
demonstrations in Brighton Beach, Brooklyn, were turning out to be a
bust. He had managed a few straight and graceless flights, but these only
proved how advanced the Wrights were by comparison. Wilbur and
Orville managed to avoid gloating.

By the end of the second week of August, Wilbur realized he would
need a larger venue to continue his demonstrations, both to accommodate the growing crowds and to allow for longer and more challenging
flights. He petitioned the French military to grant him use of Camp
d'Auvours, an artillery testing ground east of Le Mans. The request was
immediately granted. From this point on, Wilbur Wright would be
denied nothing in France.

"You never saw anything like the complete reversal of position that
took place," he wrote to Orville. "The French have simply become wild."
Within days of his first flight at Le Mans, he had been rechristened with
a French nickname ("The Birdman"), had become the subject of a hit
French song ("Il Vole"), and had become the most cherished American
curiosity to visit France since Benjamin Franklin. "I cannot even take a
bath," he wrote to his sister, "without having a hundred or two people
peeking at me."

Wilbur continued to stun spectators over the next several days. He
threaded between trees, rose above the canopy to seventy feet, then

slipped gently back to earth. He demonstrated a command of the air that had been, until this moment, unimaginable to those who saw it. On Thursday, August 13, he made his last flight at the Hunaudières race-track, circling seven times and remaining aloft for eight minutes. And he was only just starting to fly.

Springfield

Many hours later, several turns of the globe to the west, the day that had begun so spectacularly in Le Mans, France, came to a close in the small American city of Springfield, Illinois. As midnight of August 13 approached, most of Springfield's residents were asleep, thousands of Little Nemos lost in their respective Slumberlands. Perhaps they dreamt themselves flying over the city in an airship or shivering next to an American flag at the North Pole; or saw themselves standing in crisp uniform on the deck of a great white battleship, sailing toward an unknown fate in the Orient; or steering an automobile into a cheering throng of Parisians. Reality provided a surplus of fantastical images to furnish dreams in 1908. Unfortunately, it also provided ample fodder for nightmares.

Located near the center of Illinois, Springfield was the state's capital. In most ways it was a typical northern industrial city, its economy fueled by the local coal mines and the railroads that crossed through the city. Other industries included shoe manufacturers, grist and flour mills, and the Illinois Watch Company, a supplier of precise time instruments for the railroads. Together, these industries supported a population of forty-seven thousand, including about three thousand African Americans.

Springfield's most famous product, and greatest source of pride, was the sixteenth president of the United States. Abraham Lincoln had ridden into Springfield as a young man in 1837, all his worldly possessions crammed into the saddlebags hanging from his horse. In Springfield, he'd established himself as an attorney, launched his political career, married his wife, and raised his family. The city had grown with Lincoln—and partly because of him—from a small prairie settlement into a bustling metropolis.

Lincoln's presence still hovered over the city in 1908. His home stood on the corner of Eighth and Jackson Streets, a few blocks from the courthouse, and his remains lay in nearby Oak Ridge Cemetery, buried ten feet under a towering granite obelisk. During the next forty-eight hours, the Great Emancipator would have numerous opportunities to roll over in his grave.

The night of August 13 was warm and still in Springfield. A pretty young white woman named Mabel Hallam had gone to sleep early in her small home on North Fifth Street. Her husband, Earl, was working the night shift as a streetcar conductor. At about eleven-thirty, she later told police, she was woken up by the weight of a man's body upon her. "Why, Earl, what is wrong with you?"

"I am drunk," the man responded.

Mrs. Hallam realized that the man was not Earl. It was not like Earl to throw himself upon her so roughly, she later told reporters. "My husband does not possess such habits." Moreover, she realized with alarm, the man on top of her was black. Before she could scream out, the man covered her mouth with his hand and told her he would kill her if she made noise. Whereupon he raped her and fled into the night.

The next morning, Friday, August 14, Springfield residents were greeted by news of Mrs. Hallam's assault in bold headlines. NEGRO'S HEINOUS CRIME, screamed the *Illinois State Journal*. DRAGGED FROM HER BED AND OUTRAGED BY NEGRO, blared the *Illinois State Register*. The story of the assault was shocking enough in its own right. But to white residents of Springfield it was doubly egregious. Just five weeks earlier, another black man, a drifter named Joe James, had been arrested for assaulting a young white girl and killing her father in a scuffle.

As Springfield digested this newest assault, police scoured the Hallams' neighborhood for black men. Throughout the day, they marched men to the Hallam house for Mrs. Hallam to view. At last, she picked one as her assailant. His name was George Richardson. He was a laborer, a hod carrier, married but childless. Newspapers would later suggest that he was an ex-convict who had done time for murder. In fact, he was a hardworking young man with no criminal record. All that really mattered now, though, was that Mrs. Hallam's accusation made him a black suspect in the rape of a white woman. George Richardson would have been the first to understand that his prospects for an early death had just risen considerably.

By 4 P.M., a mob of whites was gathered at the county jail, shouting racist epithets and clamoring for the release of both George Richardson and Joe James into its custody. What the whites intended to do to these two black prisoners did not, in 1908, require much imagination.

Putting aside for a moment the fact that this mob was assembling in a northern industrial city that happened to be Abraham Lincoln's hometown, there was nothing unusual about whites thronging in front of a jailhouse demanding vengeance on a black man. The scene was reprised throughout America dozens of times a year in the early part of the twentieth century. Typically, as in Springfield, the inciting incident was an accusation of rape leveled by a white woman against a black man. To judge from newspapers of the time, black-on-white rape had reached epidemic proportions in 1908. In retrospect, it's clear that a good portion of the accusations against black men were exaggerated or fabricated, but once made, they were unlikely to be seriously challenged. The accused black man ran a significant risk of never making it to court.

The summer of 1908 had already seen numerous lynchings of black men. Several of the lynchings had been grotesquely sadistic even by the usual standards. At the end of July, in the town of Greenville, Texas, a crowd had forced police to hand over a young black teenager charged with assaulting a young white woman. Before hanging the boy, the mob poured kerosene over him and burned him alive. Two days after that, four black men—three of them brothers—were lynched by a mob in Russellville, Kentucky. Two days after that, in Norfolk, Virginia, three hundred white sailors tried to lynch three black men, failing only because police managed to hurry the black men out of town. Altogether, eighty-nine blacks would be lynched before the year was done, the most in any year of the twentieth century after 1901 (when 105 blacks were lynched).

Not surprisingly, most of the lynchings occurred in the south. Southern states had spent the last two decades systematically dismantling virtually every right won by blacks in the Civil War. The Fourteenth Amendment to the Constitution guaranteed African Americans equal protection under the law, but Jim Crow statutes segregated them to inferior working and living conditions. The Fifteenth Amendment guaranteed suffrage to black men, but southern states effectively imposed political disenfranchisement by means of poll taxes, literacy tests, and other obstacles to black voters.

In addition to these quasi-legal tactics of subjugation, southern whites imposed a reign of physical terror over blacks that was capricious and ruthless. The standing threat of lynching, along with other types of physical violence, made life in the south so intolerable that many blacks were leaving their homes and moving north, pioneers in the Great Migration that would continue for the next several decades. The indus- trialized cities of the north had far more to offer in the way of jobs and economic opportunity to African Americans, but this hardly made them hospitable. "[T]he more I see of conditions North and South," the cele- brated American journalist Ray Stannard Baker wrote, "the more I see that human nature north of Mason and Dixon's line is not different from human nature south of the line."

Baker was concluding a two-year investigation of race relations around the country. In his 1908 book, *Following the Color Line,* he observed that as long as blacks represented only a trifling percentage of the population, northerners tended to treat them civilly. But as blacks increased in num- bers, the white north began to eerily resemble the white south. Many employers refused to hire blacks. Labor unions refused to admit them. Landlords refused to rent to them, or did so at exorbitant rates. In some northern cities, like Indianapolis, "bungaloo gangs" of white toughs prowled the streets, randomly pouncing on blacks. In some smaller northern towns, local laws prohibited black residency. Even in Boston, for- mer seat of the abolition movement, virulent racism had taken root. "A few years ago no hotel or restaurant refused Negro guests," Baker wrote after visiting that city; "now several hotels, restaurants, and especially con- fectionary stores, will not serve Negroes, even the best of them."

Springfield had seen plenty of deterioration in its own race relations in the twentieth century. Following the arrival of the first black man in the 1830s—a barber named William Florville who came, according to lore, at the suggestion of Abraham Lincoln—blacks had prospered in Springfield. Many, including doctors and newspaper editors, had enjoyed middle-class affluence and respectability among whites as well as blacks. In the coal mines, working-class blacks and white immigrants had worked together seamlessly. Recent years, though, had seen black resi- dency increase. And with it the old racial amity had started to dissolve.

The greatest surprise to Ray Stannard Baker was not the contempt working-class whites felt for blacks, which he could attribute to compe- tition for jobs. Rather, it was the attitude among educated and more pro- gressive white northerners, the same enlightened class that had been so instrumental in the abolition movement. "A few of the older people still preserve something of the war-time sentiment for the Negro; but the people one ordinarily meets don't know anything about the Negro, don't discuss him, and don't care about him."

White northern elites did not generally condone violence or political disenfranchisement against blacks. Most, though, believed blacks incapable of achieving equal status with whites. Education had made these whites more broad-minded in some respects, but it also filled them with pseudoscientific theories of racial hierarchy. Their dismissal of blacks was intellectual but no less pernicious for it. Even Ray Stannard Baker, a progressive and a man of high morals, could not help drawing distinctions between the "best negroes" and the lower kind. The best negroes, by this reckoning, were inevitably the light-skinned mulattoes. In 1908, whites—even the best whites—had an exceedingly difficult time distinguishing between Caucasian blood and moral worth.

This left African Americans with few friends and difficult choices. Booker T. Washington counseled his fellow blacks to practice patience and to live beyond reproach in a nonconfrontational relationship with whites, and eventually justice and respect would come. Younger black leaders like W.E.B. DuBois, a brilliant Harvard-educated scholar, had recently started to question Washington's tactics of "accommodation." Equal rights were their due, argued DuBois, not something black people had to earn. DuBois's view would become the foundation of the American civil rights movement later in the twentieth century.

In the meantime, though, African Americans had to survive the present.

The mob in front of the county jail swelled throughout the evening of August 14, more clamorous and menacing by the minute. At least four thousand people had joined the crowd by 5 P.M. Sheriff Charles Werner, belatedly realizing that he had a dangerous situation on his hands, contacted the governor to request assistance from the state militia. As he waited for troops to mobilize, the sheriff hatched a plan to remove George Richardson and Joe James from the prison. Once they were gone, he hoped, the crowd would disperse and go home.

Sheriff Werner telephoned Harry Loper, one of the town's leading citizens, for assistance. Loper was the owner of the largest restaurant in town. More pertinently, he was the owner of a large, fast automobile, which he now agreed to lend to the cause of transporting the prisoners from town. Sheriff Werner had the fire department respond to a false alarm to distract the crowd. As fire trucks raced down the street in front of the prison, Harry Loper drove into an alley in back. The two prisoners were quickly loaded into his automobile, alongside two deputies. Loper sped off, heading north out of the city.

As soon as the prisoners were safely out of reach, the sheriff went to the front steps of the prison to inform the crowd they were gone. He was greeted by violent skepticism. "Lynch the nigger," someone shouted. "Break down the jail!" The sheriff invited a small delegation to inspect the building and see for themselves. The delegates searched the cells, then returned outside to confirm that the prisoners indeed appeared to be absent. Contrary to the sheriff's expectations, this bit of intelligence did nothing to placate the mob; it only removed an immediate focus for its wrath. When word passed through that Harry Loper had helped in the escape—apparently, one of the sheriff's deputies had leaked this—a cry went up: "On to Loper's!"

Loper's restaurant was three blocks from the courthouse. By the time the mob arrived, Harry Loper himself was inside with several employees, back from depositing the prisoners and deputies at the outskirts of the city. His automobile was parked on the street in front of the restaurant. It was upon this that the mob first pounced, turning it over, belly up. This done, people began throwing bricks and bottles at the front window of the restaurant. Huddled inside with his terrified staff, Loper was armed with a rifle but reluctant to shoot, no doubt hoping the police would come to his aid as he had come to theirs. He waited for them in vain. What little police protection the sheriff sent his way—a total of four officers—was useless in the face of the onslaught. Rather than risk their lives, the four officers stood back and let the mob do its work.

The destruction was only just getting under way when a plump woman named Kate Howard stepped forward to take charge. A lodging-house proprietor, she now became the mob's impassioned Joan of Arc. "What the hell are you fellows afraid of?" she goaded the men in the mob. "Come on, I will show you fellows how to do it." Following her, the mob stormed the restaurant, smashing mirrors and furniture, shredding drapery and raiding the liquor stock. Loper and his employees scrambled out through a basement window.

Back outside, at around 9 P.M., the mob stacked furniture and fixtures from the restaurant on top of Loper's overturned automobile. Somebody dropped a match into the car's gas tank and the car exploded, igniting the first of many fires that night. Many in the crowd were drunk by now, fueled by liquor from Loper's, and they began to dance in "frenzied delight" around the fire, chanting murderous slogans. "Curse the day that Lincoln freed the nigger," somebody shouted. "Abe Lincoln brought them to Springfield and we will run them out."

That is exactly what the mob attempted to do over the next six hours. From Loper's, it poured into a nearby area of town known as the Levee, where many of Springfield's African Americans lived and worked. Led by Kate Howard and several men, the crowd advanced through the Levee, then onward to another black neighborhood known as the Badlands, shooting up, smashing, looting, then setting fire to homes and businesses. When black men appeared, rioters chased them down and beat them. When firefighters arrived to extinguish the blazes, they blocked the fire wagon's access or cut the fire hoses. White residents who feared their homes might be inadvertently destroyed draped white sheets from their doors to advertise their race.

Many of Springfield's blacks had already fled the city by the time the mob arrived in the Badlands. Some had gone by streetcar, others hurrying away on foot, carrying whatever belongings they could into the

The burned district of Springfield, Illinois, probably on August 15, as whites tour the damage.

nearby countryside to spend the night in the woods or fields. The blacks
who remained in the city were in many cases too old or infirm to flee,
but the mob took no pity. Coming upon an elderly black man named
Harrison West, the mob beat him severely. Finding a paralyzed man
named William Smith, it dragged him from his house, beat him, then
tossed him into a patch of weeds.

At 2 A.M., the mob arrived at the home of Scott Burton, a fifty-six-
year-old black barber. Burton had sent his wife and family away but had
stayed behind to defend his property. As rioters approached, Burton
fired buckshot, but this did little to impede them. He tried to escape
through a side door of his house but was overtaken and knocked unconscious. In the light of burning buildings, the rioters lynched Burton
from a nearby tree, slashing his clothes and flesh with knives. They were
still mutilating his corpse when a line of state militia approached and
fired over their heads. They scattered into the night.

The following morning, the city smoldered quietly. Throughout the
early hours, militia had been arriving from around the state on special
trains, and soldiers now patrolled downtown. Curious sightseers and
amateur photographers congregated at the scenes of the riots. White
women in white summer dresses strolled the Negro districts to view the
destruction as young men carved souvenir splinters from the tree where
Scott Burton was lynched. By the end of the day, the tree would be
gone entirely.

The mayor offered refuge in the state arsenal to those blacks who had
remained in town and feared a reprisal. Some took him up on the offer
but many chose to join the exodus that had begun the previous night. All
day, a constant stream of black citizens walked out of the city. "Tears
streamed down the face of many a negro as he hobbled along on a cane
trying to assist his wife to flee to a haven of refuge outside the city,"
wrote the reporter for the *Chicago Tribune*. Now and then a white person
would call to someone he recognized. "You're not leaving town are you,
Sam? Stay, nobody will harm you." The black people just kept walking,
even though they had no place to go but the fields and woods beyond
the city. Many of the towns within a twenty-mile radius were warning
them to stay away. In the town of Buffalo, fifteen miles from Springfield,
a large sign threatened them with death. "All niggers are warned out of
town by Monday, 12—sharp. The Buffalo Sharp Shooters."

The riot resumed in Springfield that evening. Shortly before sundown, a crowd of whites began to gather near the courthouse. Smaller mobs spontaneously appeared throughout downtown. More than a thousand whites joined a press outside the arsenal, apparently working up their nerve to storm it and attack the blacks inside. They were deterred only by the armed militia posted outside.

At about eight, a group of five hundred club- and gun-wielding riot-ers appeared at the home of eighty-year-old William Donnegan. A retired cobbler, Donnegan lived in a neighborhood of mostly middle-class whites. He had prospered as a highly respected citizen of Springfield for years, and had been friendly with Abraham Lincoln as a young man. Donnegan's main offense seems to have been marrying a white woman. As a measure of how secure he felt in the city he'd lived in since 1845, he'd refused to flee when the riots began. He was so racked by rheumatism that fleeing probably would have been impossible in any case.

As the rioters approached, Donnegan stood in front of his house, as if awaiting guests. "Good evening, gentlemen," he greeted them cordially. "What can I do for you?" He was immediately hit with a brick and knocked to the ground. The crowd fell upon him. Some beat him, others cut him with a razor. A clothesline was snatched from a neighbor's yard and lashed around his neck. The rioters dragged him to a small maple tree in front of the school across the street. They tried to hoist him, but neither the rope nor tree was strong enough to bear his weight. He was still alive, lying on the ground with the clothesline around his neck, when troops rushed in. They were too late to save him. William Don-negan died later that night.

On Sunday morning, a white journalist named William English Walling arrived in Springfield by train from Chicago. A former hand at Jane Addams's Hull House, a certified progressive and committed socialist, Walling now contributed to *The Independent*, a left-leaning journal with a large circulation. Walling and his wife spent a sweltering Sunday tour-ing the city and interviewing the citizens of Springfield, attempting to piece together how the events of the weekend could have occurred in the same northern city that produced Abraham Lincoln. Walling was stunned by the lack of remorse among the whites as they strolled the streets after church. Many of those he encountered, even those who regretted the violence, admitted they were pleased that blacks had been driven from the city. "Springfield," Walling would write in *The Independent*, "had no shame."

Nor, apparently, did the rest of the northern United States. "If these outrages had happened thirty years ago, when memories of Lincoln, Garrison and Wendell Phillips were still fresh," Walling asked his readers to consider, "what would not have happened in the north?" What happened in the North now was a general sense of outrage, followed by tenuous explanations, followed quickly by nothing more.

In two nights of rioting, two blacks had been lynched. Hundreds more had been injured, some critically, by gunshot wounds or contusions from bricks the rioters had clawed up from the sidewalks and used as missiles. More than forty black homes had been destroyed by fire, twenty-one black businesses damaged, and the great majority of black residents forced to flee the city.

Authorities promised swift justice after the riot, but the response turned out to be more swift than just. Of the thousands who participated, only a hundred whites were indicted. It became clear as soon as the trials began that white juries would refuse to convict the rioters, no matter how compelling the evidence. One participant did receive a twenty-five-dollar fine for petty larceny and another, a teenager, was sent to a state reformatory, but that was about it. Kate Howard, the Joan of Arc who inspired the crowd that first night, committed suicide in late August rather than face jail. In all likelihood, she would not have ended up there.

Predictably, Springfield's blacks suffered the greater consequences. The city government dismissed almost fifty black civil servants in the immediate aftermath of the riot, including four police officers and five firefighters. The mayor feared that their continued employment would inflame whites. After the state militia pulled out on August 19, those blacks who remained in Springfield were frequently attacked by gangs of whites who rallied to continue the pogrom. This went on for weeks.

Joe James was executed for murder. George Richardson, the young black man whose alleged crime had precipitated all of this, was set free after Mabel Hallam admitted she had fabricated the rape to cover up an adulterous affair. The truth was that her boyfriend had hit her; she'd invented a savage black man to explain the bruises to her husband.

Mrs. Hallam moved away from Springfield with her husband. George Richardson returned home to his wife, and to his city. So, too, did many of the roughly two thousand blacks who had fled during the riots. It was, after all, their home.

For most white Americans, the riot quickly faded into the recesses of memory. But not for William English Walling, and not for many black Americans. For them, the nightmare of Springfield was a wake-up call—a clanging bell that could not be ignored.

A Near Miss

President Roosevelt was quiet during the Springfield riot. He left it to Taft, his successor, to publicly condemn it. Roosevelt surely condemned it, too, privately. Not because he was the second coming of the Great Emancipator, as blacks had once expected him to be (he'd disillusioned them long since), but because he was a law-and-order man who could not abide riots, whatever their cause.

The summer had been a busy one for Roosevelt at Oyster Bay. Having gotten Taft nominated, he now applied his energies to getting him elected in November. Roosevelt's staff was said to be so overworked carrying out his political machinations that many had applied for sick leave. The president himself dashed off quick notes to the candidate regarding matters of political etiquette. No more playing golf in public, he advised; it gave the wrong impression. Better to keep a rich man's sport out of sight of the electorate, as he, Roosevelt, had done with tennis.

There were signs coming out of Taft's camp that the nominee was getting fed up with his mentor's meddling. None of Roosevelt's cabinet had been invited to meet with Taft at his Hot Springs retreat over the summer, a fact newspapers interpreted as a "silent boycott" of the Roosevelt White House. This was the first indication that Roosevelt's relationship with Taft would not be quite as chummy in the future as in the past.

On August 20, Roosevelt was playing tennis on his court at Sagamore Hill when the Thomas Flyer motored up the drive. Seeing the automobile from the tennis court, Roosevelt greeted the racers with a wave of his racquet and pointed them to the house. A few minutes later, as George Schuster, George Miller, and their old teammate Montague Roberts looked around the estate, the president appeared from over a hill, hollering greetings. He invited them into his library for cigars. When all were seated, the president congratulated the men on their triumph and interrogated them about their journey across Manchuria and Siberia.

Schuster, who apparently knew his audience's narrative tastes, thrilled the president with tales of Manchurian brigands, wolves, and tigers.

"Were you well armed, Mr. Schuster?" the president inquired.

"Yes, Mr. President, we had express rifles, Colt revolvers, and shotguns."

The president laughed. "Sounds like my Africa trip, doesn't it?"

According to a *New York Times* reporter who was present, Mr. Roosevelt went on to tell his guests "that he admired Americans who did things, whether it was up in an airship, down in a submarine, or in an automobile." The president then took the party into an oak-paneled room where they could inspect the numerous stag and bear heads mounted on his walls. He assured the men that he had killed each of the animals himself.

Two days after the visit by the crew of the Thomas Flyer, the *New York Times* printed an excerpt from an interview with the president by Ray Stannard Baker; the interview, conducted earlier in the summer, was to be published whole in the September issue of *The American* magazine. Baker had found the president in an uncharacteristically wistful mood. "Well, I'm through now," the president had said quietly. "I've done my work." He still had another half year to serve but felt his end nearing as president. His thoughts strayed more than ever to Africa. "The best thing I can do is go entirely away for a year or more—out of reach of everything here; and that is what I am going to do." When Baker suggested to Roosevelt that the public might not be quite through with him—and four years from now might want him back—Roosevelt dismissed the idea. "No," said the president, "revolutions don't go backwards." Several times he simply repeated the phrase "I'm through now."

He was closer to being through than he realized. Just past midnight, August 24, as Roosevelt slumbered in his bed, a small meteorite bored into earth's atmosphere and blazed across a clear starlit sky toward Oyster Bay. Secret Service agents who were posted outside heard a hiss, then saw a flash that bathed Sagamore Hill and its lawn and tennis court in intense light. The meteorite landed less than a hundred yards from the house, just missing the president of the United States. It exploded into a hundred fiery pieces and disintegrated into dust.

CHAPTER NINE

A Month of Late-Summer Days

AUGUST 21–SEPTEMBER 21

Within the next thirty days two continents will be ringing with the news of these performances, for when we look at the matter soberly there is no enterprise before the world to-day of greater moment than this new business of air travel.
—Byron R. Newton
Van Norden Magazine, August 1908

The bloodshed of Springfield, disquieting as it was to many Americans, did not trouble the nation's psyche for long. By September, newspapers and politicians had moved on to other matters; and by the time Lincoln's centennial came around the following February, the riot had been pretty well demoted to a footnote in history, a place it has held ever since. This early-onset forgetfulness was perhaps a testament to how accustomed whites had become to violence against blacks; and perhaps, also, a measure of how many privately condoned what they publicly condemned. Black Americans did not have the luxury of forgetting Springfield. But for whites, the whole sorry mess was best swept under the rug.

There was another reason to forget. Two reasons, in fact, both named Wright. Between the third week of August and the third week of September—a month of late-summer days that immediately followed the Springfield riot—Orville and Wilbur put on an intercontinental display of aerial prowess that was awesome and transcendent. As the brothers sailed through the sky, one in France, the other in America, they wiped troubled minds clean.

Wilbur resumed his flights in Le Mans on August 21. He had moved his base of operations to Camp d'Auvours. Enormous crowds came rushing from all over Europe, traveling by automobile or by train, to see him fly. Many who lived in the French countryside simply rode their bicycles, pedaling as far as twenty or thirty miles to be there. Within the town of Le Mans itself, tourism increased markedly. Cafés bustled and money flew. The higher Wilbur Wright rose in the sky, it was said, the higher prices rose at Le Mans hotels.

Wilbur had erected a new, larger shed at Camp d'Auvours. For the next several months he would work, eat, and sleep in this shed, never leaving his airplane's side. Most likely, Wilbur bedded down next to the plane for practical and proprietary reasons—it was the best way to protect the machine from harm or prying eyes—but to the French, his domestic arrangements suggested an abiding devotion to his calling and became another feature of his emerging mystique.

Wilbur Wright captivated the French. American newspapers tended to concentrate on the details of the flying machine the Wrights had invented; French journalists studied the man himself. They were intrigued by Wilbur's quiet resolve and spartan self-sufficiency. They conceived of him as a kind of poet/mystic, like "the monks of Asia Minor who lived perched on the tops of inaccessible mountain peaks," one writer put it. Other than his countenance, inevitably described as birdlike, it was Wright's eyes that enchanted them. One journalist described the blue-gray orbs as "courageous, gentle, determined, intelligent." Another saw in them "the crystal clear mirror of his beautiful, pure soul." Some of this was no doubt writerly excess, but the descriptions also reflected French pride. It was not enough that the man who bested them be merely a man; he must be a seraph or saint.

Drawn as much by the instant legend of Wilbur Wright as by the wonder of flight itself, crowds of would-be spectators descended on Camp d'Auvours throughout late August. The great majority went back home having witnessed nothing more interesting than the exterior of Wright's shed, for on most days Wilbur did not fly. Sometimes artillery practice made the range inaccessible. Often it was blustery or rainy weather, or mechanical difficulties, that prevented the machine from rising. And sometimes it was not entirely clear what kept Wilbur on the ground. The wind would die, the artillery fire would cease, but Wilbur would remain in his shed, working and whistling. The crowds waited, and Wilbur let them. Like spurned lovers, they responded with only more ardor. What sort of man was this who ignored the conventions of courtesy? A man who would make thousands wait while he whistled! "Even if this man sometimes deigns to smile, one can say with certainty that he has never known the *douceur* of tears," the aviator Léon Delagrange wrote. "Has he a heart? Has he loved? Has he suffered? An enigma, a mystery."

In fact, behind that sphinxlike surface Wilbur was more agitated than he had been before his trials began. His success in France had been terrific, but success had only brought him worry. He worried that the success was *too* great, that the overblown accolades would give rise to a backlash. (He was right.) He worried that Orville would meet trouble at his impending trials in Washington. (Right again.) He even worried—despite the impression he gave to the contrary—that he was disappointing the crowds by refusing to fly. "[W]hen I think of the sacrifices some of them have made in the hope of seeing a flight," he wrote in a letter to his father, "I cannot help feeling sorry for them when I do not go out."

Whatever sympathy he felt for the spectators, he also despised them a little. Their presence cramped and sapped him. All day people surrounded his shed, trying to peer in at him through the cracks. When the shed door opened, they stared at him dumbly, making him feel as if he had no room to breathe. "I can't stand it to have people continually watching me," he wrote to his father. "It gets on my nerves." The constant attention, he later explained, "had brought me to such a nervous point of exhaustion that I did not feel myself really fit to get on the machine."

And so they waited, and he waited. They were waiting for him to fly. What he was waiting for, really, was for them to go away so he could practice in peace.

But, of course, it was much too late for that.

On August 22, the day after Wilbur returned to the sky over Le Mans, his younger brother, Orville, got his first look at the grounds of Fort Myer, Virginia. Orville had arrived in Washington, D.C., from Dayton on the evening of August 20. Charlie Furnas and Charlie Taylor, both good mechanics and trusted friends—"the two Charlies," the Wrights called them—were already in Washington to lend a hand. The airplane had been shipped in crates from Dayton and now awaited Orville inside a tent at the edge of the Fort Myer parade grounds. The Charlies had already begun to assemble it.

In many ways, Orville's trials at Fort Myer, occurring in his home country and facilitated by trusted allies like Furnas and Taylor, promised to be
a smoother ride than Wilbur's in France. But he would face his own dis-
tinct hurdles. According to the specifications of the Wrights' contract with
the army, Orville was required to successfully complete a series of flights
that would push the plane—and Orville's flying ability—beyond anything
either machine or man had attempted. One of the trials, an endurance
run, required Orville to remain in the air for at least one hour, almost twice
his personal best. In a speed test, he would have to achieve an average
velocity of forty miles per hour over a five-mile cross-country circuit,
which would mean pushing the plane's motor to the brink of its capacity.
He would also have to prove that the plane could carry a passenger and
that its operation was simple enough to allow "an intelligent man to
become proficient in its use within a reasonable length of time." A panel
of five judges would observe his flying to determine whether he met the
army's criteria.

When the conditions of the contract had been announced eight
months earlier, many of America's aviation experts had denounced them
as impossibly demanding. No machine at that time had *publicly* flown so
much as an inch in America. Some of that initial skepticism had faded
by late summer of 1908, following the widely reported flights of the
previous months, but the prospect of a heavier-than-air machine
remaining aloft for an hour still seemed, to many, remote.

As Orville toured Fort Myer that day in late August, a balloonist
named Thomas Baldwin, also competing for a contract with the Signal
Corps, had just successfully completed flight trials in his sixty-six-
foot-long dirigible. Another competitor, Augustus Herring, had called
the Signal Corps from New York City to request a one-month extension on the delivery of his own heavier-than-air machine. Herring
promised to fly down from New York when his machine was ready.
Orville was not the only one who found this boast a few hundred miles
shy of plausible.

Orville toured the grounds of Fort Myer in a government-issue automobile with two officers of the Signal Corps. The three men drove on a
long course around the military base so that Orville might see it in full.
With its high promontories, Fort Myer had much to recommend it as a
headquarters for the Signal Corps. As a proving ground for an airplane,
however, it left a lot to be desired. The terrain was uneven and rugged,

rising from sea level to 240 feet and including several deep ravines and large woods. The Wrights had scrupulously avoided flying over this kind of terrain in the past because it offered few opportunities for bailouts or crash landings. For the cross-country speed test, though, Orville would have no choice but to fly over it and take his chances.

Orville read the land as an aeronaut, seeing in its rises and hollows not just topographical features but also atmospheric implications. He believed that trees and buildings, for instance, influenced air pressure negatively, pulling the plane downward. The oaks of Arlington Cemetery, at the eastern edge of the parade ground, gave shade to the graves of the Union soldiers buried under them; but to Orville, they represented a threat from below. ("Whenever I come near the cemetery where all those high trees are the machine has a tendency to come down," he told reporters after he started flying. "I don't want to go to the graveyard just yet.") Air was invisible but alive. It flowed and rippled over every innocuous-looking feature of the earth. Orville would have to be ever vigilant to its ways.

And there was an added challenge: Orville was on his own. He was not accustomed to this. Wilbur had always played the older brother, both in public and private, as a function not only of primogeniture but of temperament. Though Orville was lively and funny in small groups, he was diffident in public settings and had always deferred to Wilbur. He would not be able to do that at Fort Myer. He alone would represent the brothers to the Washington press corps, to Washington society, to prestigious government officials and the United States Army—and to the world. He would have friends there to help, but he would have enemies, too. Indeed, in some ways, the crowd at Fort Myer would be tougher than the one Wilbur faced in France. Several members of the AEA would be present, including Glenn Hammond Curtiss, the man who had flown the *June Bug* on July 4 (and who would be the Wrights' nemesis for years to come). Another AEA member, Lieutenant Thomas Selfridge—the same Thomas Selfridge who had written the Wrights back in January to solicit advice on building a plane—had been appointed as one of the five judges overseeing the trials. Selfridge was generally a well-liked man, tall, handsome, and intelligent, but Orville developed an instant antipathy for him. "I don't trust him an inch," he wrote in a letter to his brother. "He is intensely interested in the subject, and plans to meet me often at dinners, etc. where he can try to pump me ... I understand he does a good deal of knocking behind my back."

In words that would later appear chillingly prophetic, Orville also wrote, "I will be glad to have Selfridge out of the way."

For the most part, Orville found Washington open and welcoming. By arrangement of a well-connected old acquaintance, Albert Zahm, he moved from his hotel into the tonier surroundings of the Cosmos Club on Lafayette Square. The members of the Cosmos took a shine to him, as did many others. Pulled into the orbit of the capital's elite—that portion of it, anyway, that could be found in the torpors of a late August Washington—Orville was dined and honored about town. He found himself enjoying his role as sole representative of the Wright franchise; indeed, he turned out to have a knack for it. Like Wilbur, he had a winning ability to communicate humility and conviction in equal measure. "The reporters seem to think I am not in the least uneasy about fulfilling our contract," he wrote to his sister, Katherine, on August 31. "They say that I do no boasting of what I can do; that they can get but little out of me as to what I expect to accomplish, but that I have the air of perfect confidence!" When not handling reporters, Orville also got a chance to indulge in the attention of women. "I am meeting some very handsome young ladies!" he informed his sister. "I will have an awful time trying to think of their names if I meet them again."

Society, reporters, women—these were exactly the kinds of distractions that Wilbur worried his less ascetic brother might face in Washington. "I fear he will have trouble with over-attention from reporters, visitors . . . &c.," Wilbur wrote to their father. "It is an awful nuisance to be disturbed when there is experimenting and practicing to be done."

True to Wilbur's fears, Orville found little time to work. "The trouble here is you can't find a minute to be alone," he wrote to Katherine. "I have to give my time to answering the ten thousand fool questions people ask about the machine." He was also having trouble sleeping, he confided to his sister.

Wilbur's well-meaning letters probably did little to ease Orville's slumbers. They were littered with apprehensions and admonishments. "I advise you most earnestly to stick to calms till after you are sure of yourself," Wilbur wrote on August 25. "Do not let yourself be forced into anything until you are ready." Later in the same letter, Wilbur wrote, "Do not let people talk to you all day and all night. It will wear you out before you are ready for real business . . . I can only say be extraordinarily cautious."

On August 29, Wilbur wrote, "Be careful of your electrical connections." And on August 30, "Be exceedingly cautious as to wind conditions and thorough in your preparations. . . . I wish I could be at home."

On the morning of September 3, during a break in the rain, Wilbur flew in Le Mans, a jaunt of ten minutes and forty seconds. Later that same day—it was morning now in Washington—Orville left the Cosmos Club and boarded a streetcar to Fort Myer. He had let it be known that he intended to fly.

The day began clear but windy in Washington. As afternoon descended, though, the wind gradually died. Conditions were calm by 5 P.M., when Orville asked his assistants to remove the plane from the balloon shed at the southern end of the parade ground. Newspaper reporters and photographers followed as Signal Corps men carried the plane to the launch rail across the field. Orville, dressed in his usual gray suit, high starched collar, and Scotch plaid cap, led the men through every step of the setup. "For the first time since he arrived in this city he betrayed obvious signs of nervousness," discerned the reporter for the *Washington Post.* "The lines in his face seemed deeper, and there was an uneasiness of manner which was noticeable."

Several hundred spectators were present to witness Orville's inaugural flight, including the high command of the Signal Corps, several members of the president's cabinet, and the president's eldest son, twenty-one-year-old Theodore Roosevelt Jr. Men stood about in boaters and suits, women in summer dresses and wide hats. As mounted soldiers held the crowd back from the plane, Orville and his assistants tinkered endlessly. The *Post* reporter noticed that Orville busied himself with tasks that could easily have been performed by assistants. He seemed to want to keep his nervous hands occupied.

Finally, six men hoisted the twelve hundred pounds of iron weights to the top of the catapult derrick—a sign the launch was near. Orville and Charlie Taylor gave the propellers a pull and the motor sputtered to life. Almost immediately, though, Orville cut the motor. In one version, an assistant ran to the balloon shed for a tool needed to make a small adjustment. In the *Post*'s version, it was Orville himself who went to the shed, calling out, "I'll get it myself. The walk won't hurt me any." He strode down the field and entered the dark tent.

"That man's nerves are pretty near the jumping off place," somebody muttered. "That's why he's doing errands."

A few moments later, Orville reappeared with the tool and returned to the plane. Further tinkering ensued. Restless members of the crowd began to wonder whether they had come all this way for nothing. An older man turned to an army officer next to him and wagered twenty dollars there would be no flight. No sooner did the army officer take the bet than the plane's propellers were engaged again. The motor whined. Orville climbed into his seat and set his feet on the pedal that controlled the horizontal position of the rudder. He grasped the levers that controlled the torque of the wings.

Down the field, a woman in a black dress and black bonnet, having somehow evaded the mounted sentinels of the Signal Corps, was casually strolling across the middle of the parade ground. She heard the motor of the plane and turned in surprise. "Look out there!" a soldier cried to her. The woman lifted her skirts and ran across the field.

"I'm ready," called Orville from the plane. "Let her go!"

The catapult weights dropped. The plane shot down the steel rail, gathering speed. Then it was at the end of the rail, off the ground barely, skimming along the top of the long grass, appearing as if it would go no higher—and then it did go higher, rising over the tent, banking to the left. "It was a magical moment," the *Washington Post* recounted, "and the spell of it fired the spectators with a common impulse. That impulse was to shout."

Orville circled the field once. He was starting to round it again when the plane veered toward a wooden shed near the balloon tent. The motor went dead and silent. The plane descended quickly, hitting the ground at a sharp angle and vanishing for a moment behind an explosion of dust. Reporters and photographers ran to the site of the landing. There they found Orville casually brushing the dust from his jacket, inspecting some minor damage to the runners. He blamed the hard landing on his inexperience with the new controls. He'd accidentally moved a lever the wrong way, he told reporters. Still, no real harm was done. One of the runners was broken but could easily be repaired.

Orville's first flight at Fort Myer had lasted a minute and eleven seconds. It was neither well executed nor gracefully concluded, but to the crowd, the great majority of whom had never seen one better, it was breathtaking. Everywhere hands reached in to shake Orville's. Hardened newspapermen teared up. "Not a man who saw it can forget it," wrote C. H. Claudy, reporter for the *New York Herald*. Soldiers pushed the mob back to give Orville and the plane breathing room, but not before a woman managed to reach Orville and address him in a moment of captured privacy. "I don't know how to thank you for the glory of those few seconds," she said to him. "They were rare and wonderful and to be remembered forever."

The only person disappointed in the flight that evening was Orville. He knew that both he and the plane were capable of much more. The following day, Friday, he proved it, remaining aloft for four minutes and fifteen seconds, flying steadily at about thirty-five miles per hour over three miles. "That was better than last night, and I am getting onto the machine better," he told a reporter, "but it is not what it will do when I learn it well."

Orville did not fly Saturday. In Le Mans, Wilbur, whose nervous exhaustion and abhorrence of crowds seemed to lift as his brother began to fly in Washington, made his longest public flight yet, almost twenty minutes. His control of the plane was absolute.

On Sunday, neither brother flew. But as the Wrights observed the Sabbath, Léon Delagrange, the French pilot who had been generous with his praise after Wilbur's first flight in Le Mans, took to the sky and stole their thunder. He set a new endurance record for heavier-than-air machines, flying almost half an hour at Issy-les-Moulineaux. The next day, Monday, Delagrange flew again, this time for thirty-one minutes. Privately, the Wrights had beaten this record by a good five minutes. Publicly, their newly won supremacy of the air—the aerial supremacy of the nation—was suddenly in question. Orville's flights at Fort Myer were no longer about fulfilling a contract. They were matters of family honor and national pride.

Later that Monday, Orville flew eleven minutes at Fort Myer. This wasn't a world record by a long shot, and was a minor achievement compared to flights the brothers had made at Huffman Prairie, but it was the longest public flight yet recorded in America. Reporters noticed that Orville was flying now with ease and comfort, executing narrow and tight turns and daringly rising to heights of sixty feet. From below, he looked like an athlete who had finished warming up and was ready to play the game.

A Trip to the Moon

Orville arrived early at Fort Myer on Wednesday, September 9. Shortly before 8 A.M., he ordered his assistants to remove the plane from its berth in the balloon shed and set it on the launch rail. No more than a few dozen people were present at this early hour, just a handful of men from the Signal Corps, along with Charlie Taylor and Charlie Furnas. The only reporter present was the hardworking C. H. Claudy of the *New York Herald.* Claudy was committed not only to filing two stories a day, but also to photographing Orville's trials for the *Herald* with his Graflex camera. Each evening, his negative plates were rushed by messenger to Union Station to make the six o'clock train to New York.

Orville took off a few minutes before eight-thirty. The plane shot up into the sky and began soaring around the parade ground. After eleven minutes in the air, it was clear that Orville intended to stay up a while and try to beat Delagrange's record of thirty-one minutes. The small gathering below silently watched him circle the field, the engine of the plane crescendoing, then fading. "Don't make a move, don't make a sound," Charlie Taylor cautioned the others. "If he sees any kind of demonstration, he will think he has done it and come down."

Minutes later, somebody called out, "By Jings, he's broken Delagrange's record!" Everybody present grabbed each other's hands, each man aware, according to C. H. Claudy, that he "had actually been present while aerial history was being reeled hot from the spinning wheel which made that awkward, delicate, sturdy and perfect wonder above their heads go round and round the field."

Overhead, Orville had missed the congratulatory outburst below, so had no idea he'd broken Delagrange's record. He was lost in flying. He canted into sharp corners and dipped low, skimming over the parade ground, then suddenly rose to one hundred fifty feet, high over the tents, higher than anything visible but the needle of the Washington Monument and the dome of the Capitol rising to the east, backlit by morning sun. "I wanted several times today to fly right across the fields and over the river to Washington," Orville later confessed, "but my better judgment held me back." After fifty-eight circuits of the parade ground, he landed. He had flown fifty-seven minutes and thirty-one seconds, nearly double Delagrange's record.

"Mr. Wright was not at all excited, and apparently but little elated at his feat," C. H. Claudy wrote in the *Herald*. "But there are fifty persons who saw the finish who are going around hugging themselves that an American machine yanked the record away from Europe, that it was done by an American and that they were there to see. . . ."

That evening, Orville flew again. This time, a large audience was present. As news of the morning flight spread, thousands of Washingtonians had dropped their plans for the day and made their way to Fort Myer. Among the evening crowd were three cabinet secretaries, several assistant secretaries, numerous generals, and assorted prominent personages. "Will you try to beat your own record of this morning?" a reporter asked Orville.

"Well, we will see about that," he replied with a smile.

Orville took off at a quarter past five. The plane rose swiftly above the parade ground and banked into its first turn. This time, Orville was furnished with a watch so he would know how long he stayed aloft. The crowd below watched in near silence as the plane droned around and

Orville flying over Fort Myer, September 1908.

around above them. Minutes passed as a series of ellipses in the sky. Each circuit of the field equaled approximately the full revolution of a second hand around a timepiece. After about forty-five circuits—three quarters of an hour—some in the crowd consulted watches and began to count out the minutes. At the one-hour mark, a cheer went up. Orville took another three trips around the parade ground, then descended. He had flown for one hour, two minutes, and fifteen seconds.

The first man to congratulate Orville after he landed was the despised Lieutenant Selfridge. Then came the cabinet officers, the generals, the reporters, and the rest of the crowd, calling out three cheers in Orville's honor. Orville took it all in stride. Whatever nerves he had displayed a few days ago were gone. Without pausing to celebrate, he asked his men to set the plane back on the steel rail. Having broken one record, he would now break another. He turned to Lieutenant Frank Lahm, a member of the Aeronautic Board whom he particularly liked. "Would you like to go up with me, Lieutenant?"

"Would I? You just bet I would."

"Well, climb in, then. Just sit tight and don't fidget."

Both Wilbur and Orville had taken passengers up before. And the French pilot Léon Delagrange had flown with Henri Farman once for a few moments. But none of these flights had been publicly witnessed.

Just before seven in the evening, Orville and Lahm took their seats. The sky had darkened but an enormous harvest moon rose on cue from behind a bank of clouds. "The air was perfectly still, and there was a mysterious element of uncertainty in the silence that brooded over the scene," wrote the *Washington Post*'s reporter. Despite the weight of the extra man, the plane rose without difficulty. "As it sped along the road bordering Arlington Cemetery," reported the *New York Tribune*, "it came in a direct line between the onlookers and the full September moon, recalling to many the work of Jules Verne, 'A Trip to the Moon.'"

Six minutes and twenty-six seconds later, the plane returned gracefully to earth. Another world record had been set. "It was the most exhilarating thing I ever experienced," Lieutenant Lahm told reporters.

The following day, news of Orville's flights dominated front pages around the country and around the world. From Europe came more praise. "The conquest of the air is an accomplished fact," declared *Le Figaro*; "yesterday is a date in the history of humanity." The sentiment was echoed by Major Baden Powell, the leading expert on aeronautics in

England. "It may now be said that aerial navigation without the aid of gas is an accomplished fact." Reporters who visited the Aero Club of London seeking comment about Orville's flights were disappointed to find the place nearly deserted. Most of the members had left for France to see Wilbur.

The day after setting his endurance record, Orville returned to the sky and improved on it with a new time of sixty-five minutes, fifty-two seconds. He spent much of the flight a hundred feet above the earth. Orville had been in the air just over an hour when C. H. Claudy of the *Herald* saw "a little incident which was a convincing demonstration of what the man bird was doing."

> Another kind of bird, a dove, came winging down the field after Mr. Wright, followed him for perhaps three hundred yards in a vain effort to catch up to and examine this monster, which made such a queer noise and flew without flapping its wings, in such a constant circular path. But the effort was futile—the man bird beat the bird of the air; brains and machinery won out against instinct and perfect knowledge of flying, and Mr. Wright shot ahead, his competitor outflown, outclassed and outrun.

The next day, Friday, September 11, Orville added another five minutes to the record, this time also executing two perfect figure eights. Saturday, September 12, flying above five thousand spectators, he set three new records: seventy-four minutes and twenty-four seconds in the air flying solo; more than nine minutes with a passenger (Major Squire of the Signal Corps); and an altitude of nearly 250 feet, higher than an airplane had ever flown.

The Sunday papers were filled with photographs and descriptions of "The Wonderful Wright Aeroplane." The brothers took the Sabbath as an opportunity to rest and to write to each other. "The newspapers for several days have been full of the stories of your dandy flights," Wilbur wrote to Orville, "and whereas a week ago I was the marvel of skill now they do not hesitate to tell me that I am nothing but a 'dub' and that you are the only genuine champion skyscraper. Such is fame!"

To Wilbur, Orville confirmed that his flights were making a great impression. "The Navy Department seems much interested . . . Every

one here is very enthusiastic, and they all think the machine is going to
be of great importance in warfare."

Aerial warfare had lately become a hot topic in Washington. That
planes might be of military use was no great epiphany. After all, the
Wrights were demonstrating their plane to the Signal Corps of the
United States *Army*. But previously the military had looked to airplanes
mainly for their potential in surveillance and reconnaissance. Airplanes
could be sent over the troops of an opposing force and return with valu-
able intelligence. Unlike balloons or dirigibles, which were slow-moving,
barn-sized targets, planes were small, fast, highly maneuverable, and
presumably almost impossible to shoot out of the sky. *Scientific American*
argued in its September 12 issue that airplane scouting, by removing the
possibility of surprise attacks, would effectively end combat forever. Airplanes would make peace, not war.

It was a fine sentiment while it lasted, but it did not last long. Orville's
flights at Fort Myer inspired military minds to think beyond reconnaissance and peace. They began to see the potential of the machine as an
offensive weapon. The symbolism of that vanquished dove described by
C. H. Claudy—the bird of peace trying but failing to catch up with
Orville's "monster"—was all too apt. At a September 12 dinner of naval
officers given in honor of Orville Wright, officers discussed the ways in
which weaponized planes could be launched from the decks of boats and
flown over opposing ships. "A shell could be dropped into the funnel of
a warship," the *New York Tribune* elucidated, "causing terrible damage to
the machinery and completing its work of destruction by bursting the
boilers."

The Wrights were devout Christians but they did nothing to discourage such talk. On the contrary, they emphasized the combat applications
of their machine. The greater the potential of the airplane as a weapon,
after all, the better the Wrights' chances of selling one to a government;
and as governments were the only likely large-scale purchasers of airplanes in 1908, the Wrights honed their sales pitch accordingly.

"On your present machine, how much weight could be added in the
form of a gun?" Lieutenant Sweet, the officer assigned by the U.S. Navy
to investigate the purchase of a Wright airplane, inquired of Orville.

"One hundred and fifty pounds," replied Orville helpfully.

He went on to inform naval officers that he'd already performed
experiments in which he successfully dropped weights onto targets from

a moving airplane. "I found that after a little practice it became comparatively easy to put the weight just where I wanted to."

Apparently, this was just what the navy wanted to hear. In his Sunday letter to Wilbur, Orville informed his brother that Lieutenant Sweet was very enthusiastic. "He says he believes that every war vessel will soon be equipped with machines. They will want machines to carry two men, fly for four hours, and carry floats so that they will not sink in case the engine stops."

The following day, the *New York Times* chided Orville for emphasizing the martial efficacy of his machine over more peaceful uses. "It is to be hoped that he is wrong in this, for it would be a rather pitiful outcome if the airship so long sought should prove to be only, or even chiefly, an instrument of destruction."

Omens

The sky was more crowded than ever that September. Baseballs, skyscrapers, and now airplanes intruded on it; balloons, too, levitated into its upper

From the August 15, 1908, issue of *Harper's Weekly*: one idea
of what aerial warfare might look like.

currents, where they often burst into flame and fell like meteoroids. And then there were the *real* meteoroids that burned across the sky like ancient omens—or like missiles hurtled from the firmament to warn humans to back off. People wrote to science magazines from around the country to report sightings of peculiar luminescences. A subscriber named W. E. Ellis informed *Science* that on the night of September 4 he noted a "bright flat glow" accompanied by vertical "streamers" to the northwest of Plum Island (off the coast of Connecticut). That same night, a strange glow was observed above Lake Placid in upstate New York. Four days later, on the morning of September 8, a loud rumbling noise that sounded to some like a collision of freight trains and to others like a massive explosion of dynamite—but was probably a meteoroid entering the atmosphere—rocked central Tennessee. "It was, in fact, almost deafening," Paul Marshal of Nashville wrote to *Scientific American*, "and interspersed with three loud explosions, it crashed and roared away in—to many—a terrifying manner."

One week later, on the evening of Thursday, September 17, at a quarter past seven o'clock, a "fiery mass" passed over Massachusetts, traveling east to west. "While journeying eastwardly from the direction of Pittsfield to Springfield with a party in an automobile," a Mr. Elihu Thomson wrote to *Science*, "we were startled by a bright illumination of the landscape, like that given by an intense flash of lightning though much more prolonged. On looking upward at once for the cause, our attention was at once fixed upon a brilliant body descending rapidly and almost vertically, and apparently nearly overhead. It appeared to fall into the woods on a hill to the left of us."

If this was an omen, it came two hours too late.

Earlier that same evening, Orville flew again for the first time since his spectacular flights of the previous Saturday. Grounded by poor weather, he'd used the down time to make a minor repair to his motor. He'd also installed new propellers—"screws," the Wrights called them—with slightly longer blades. If the new propellers pleased him, he intended to use them in his official trials, which would commence in the next day or two.

The parade ground at Fort Myer had buzzed with excitement throughout the afternoon. Having been deprived of flights for several days, the crowd was large and eager to see what Orville might have up his sleeve. That morning, as some in the crowd probably knew, Wilbur had flown for thirty-nine minutes in Le Mans, his longest flight yet—and longer than anyone but Orville had flown.

At four-thirty, Orville appeared and directed his men to prepare the airplane for flight. He announced that he would begin with a two-man demonstration. Lieutenant Selfridge had asked to accompany him, and Orville had consented. Orville did not like Selfridge, but he could not very well refuse a flight to a member of the Signal Corps's Aeronautic Board.

Any personal reluctance or antipathy he felt he managed to hide. According to newspapers, both men appeared amiable and cheerful before they stepped into the plane. "Please break your nine minutes record, Mr. Wright," one of the reporters called to Orville. "Give us another record to write about." Orville laughed and promised he would do so. Then C. H. Claudy of the *Herald* called out to Selfridge: "Lieutenant Selfridge, I want you to remember when you are up there in the air that there is a man down here envying you every minute of your ride." Selfridge grinned. "I guess there will be several who envy me." The lieutenant handed his uniform jacket and hat to a friend, then took his seat next to Orville.

Orville Wright and Lieutenant Thomas Selfridge moments before takeoff on September 17. Selfridge looks into the lens of C. H. Claudy of the *New York Herald.*

The plane shot down the rail at 5:14. Spectators noticed that it rose more slowly than usual, as if struggling under the burden of Selfridge's added weight. Soon, though, the two men were high above the field, circling it. Selfridge waved to the crowd below. He could be seen talking to Orville over the motor, his arms crossed over his chest as the plane made three easy revolutions around the parade ground. On the fourth, Orville climbed to a height of a hundred feet, then banked toward the cemetery. That is when people on the ground heard a noise—"a crack like a pistol shot," one eyewitness described it—and saw something fall from the plane. The machine immediately began to turn sharply to the right.

Up in the plane, Selfridge glanced at Orville but said nothing. Not knowing what had caused the noise, or what was now causing the plane to swerve, Orville cut the engine, intending to glide to the ground without propellers. But as he worked the levers, he found the cables unresponsive. Rather than level out, the plane turned abruptly to the left, its right wing lifting. Then it turned nose down and plummeted. Selfridge uttered a quick "Oh! Oh!" but that was all.

For a second after the plane hit the ground, the crowd, watching from almost half a mile away, stood in stunned silence. The next second,

Moments after the crash, reporters and spectators work to free Orville and Lieutenant Selfridge from beneath the wreckage.

though, they began running in "a mad silent race across the field," C. H. Claudy would later write—mounted soldiers galloping toward the crash site ("a weird spectacle of my idea of a cavalry charge in actual battle," a man who was there recalled), the crowd following like foot soldiers in a massive onslaught. The first rescuers to arrive found a heap of shattered wooden spars, metal wires, and ripped canvas. Soldiers, assisted by journalists and a few physicians, tore away at the plane to reach the men buried underneath. They found Orville on his side, pale and in obvious pain but conscious. Selfridge was still alive but unconscious. He could be heard groaning softly.

By the time the men's bodies had been pulled from the wreckage, the parade ground was a scene of bedlam. The crowd pushed in to gawk and mounted soldiers pushed back, the hooves of their horses kicking up a thick cloud of orange dust over the whole scene, choking out sun and air. Physicians hurried to bandage the bodies of the injured men. Charlie Taylor leaned against the plane wreckage and sobbed. C. H. Claudy, who had come to know and like both Orville Wright and Lieutenant Selfridge personally, struggled to keep his composure and take photographs, a difficult task amid the clattering hooves and heavy dust. He glanced over to see the veteran photographer Jimmy Hare of *Collier's*—the same Jimmy Hare who been present at Kill Devil Hill in May—"taking pictures as coolly as if it were a pink tea instead of a tragedy." There was a reason, Claudy

Doctors tend to Lieutenant Selfridge.

wrote, that news photographers had recently come to be known as "Vultures of the Hour."

At the hospital, Orville remained conscious. His leg and several ribs were broken, but he was coherent enough to dictate two telegraphs, one to his sister and one to his brother, to inform them of the accident.

Lieutenant Selfridge never regained consciousness. His skull had been fractured. Surgeons removed a piece of the skull to relieve swelling, but there was nothing more to be done. Selfridge died three hours after the crash.

It was eight the following morning in Le Mans—still night in Washington—when word of the accident reached Wilbur in his shed. The news devastated him. Selfridge may have been a rival but he was one of the few people on earth who understood flight with anything like the aptitude of the Wrights. Wilbur had been planning to go up that morning, but out of respect for the lieutenant he decided to forgo flying for a few days.

The crash confirmed all of Wilbur's worst fears. He immediately attributed it to Orville's carelessness, which had been brought on, he believed, by the stresses and distractions surrounding the trials. "I cannot help thinking over and over again 'If I had been there, it would not have happened,'" he wrote to Katharine in a long letter three days after the accident.

> It was not right to leave Orville to undertake such a task alone. I do not mean that Orville was incompetent to do the work itself, but I realized he would be surrounded by thousands of people who with the most friendly intentions in the world would consume his time, exhaust his strength and keep him from having proper rest. When a man is in this condition he tends to trust more to the carefulness of others instead of doing everything and examining everything himself.

Katharine Wright quit her job as a schoolteacher and left Dayton the very night of Orville's accident. She traveled through the night and arrived in Washington the following afternoon at her brother's bedside. She would remain there for months to come, nursing him back to health.

As he lay in bed, Orville reviewed in his mind the final moments of the flight, trying to figure out what caused it. Charlie Furnas and Charlie Taylor brought him pieces of the broken plane to examine. In time, he came to conclude that a blade of one of the newly installed propellers had fractured, setting off a chain of events that led to the severing of a stay wire connected to the plane's tail. Wilbur was not entirely convinced by Orville's theory. Months later, he wrote to Orville, "I fear some questions regarding the accident can never be settled."

What the accident meant for the Wrights' future was, in its immediate aftermath, another unsettled question. Clearly, the brothers would not be able to fulfill their contract with the army any time soon. Would they ever regain the trust of the public now that one of their planes had gone down and killed a man? Many in the aeronautics community assumed that its effects would be devastating and long lasting—"the most severe setback that aeronautics has ever received," as paraphrased by the *New York Tribune*. But that gloomy assessment was almost immediately contradicted by events in the air.

Four days after the crash, on September 21, Wilbur decided the time had come to fly again. After a rainy morning and several false starts later in the day, he lifted off from Le Mans shortly after five in the evening and began circling the artillery ground at Camp d'Auvours. His largest crowd ever, ten thousand spectators, watched from below. This time, he made the trip worth their while.

"When dusk crept on and the little breeze which had been blowing died away the flight became more and more entrancing," reported the *New York Herald*, by now the unofficial chronicler of Wright triumphs and tragedies.

"From time to time the machine would disappear into the gloaming so that it almost seemed to have followed the example of the birds and gone to roost in the woods at the far end of the ground. Then would come the whirring sound of the propellers and the crackling of the motor, and out of the evening haze there appeared a slender streak of white."

First, Wilbur beat his own record of thirty-nine minutes. Then he closed in on his brother's record. People began lighting matches to consult their watches as the sky darkened and the hour mark approached. When Wilbur surpassed Orville's flight of seventy-five minutes, "a yell went up which defies description." Still, he flew. The drone of the motor came and went, and the sky grew darker and the air cooler. At last, the plane slowly descended and settled on the ground.

Four days after Orville's crash, on the evening of September 21,
Wilbur flies over Le Mans.

Wilbur had flown for ninety-one minutes and thirty-one seconds,
covering a distance of sixty-one miles. He had set a new world record
and banished any conjecture that the Wrights were finished. "I thought
of Orville all the time," he told reporters.

As for Lieutenant Thomas Selfridge, he had the dubious distinction
of being the first American, and the first passenger *anywhere*, to die in an
airplane. His body is buried close to where he fell, under the trees of
Arlington National Cemetery.

PART III

The Golden Age

CHAPTER TEN

The Certainty of the Future

SEPTEMBER 23–OCTOBER 8

I will build a motorcar for the great multitude.
—Henry Ford

The autumnal equinox fell on September 23. The earth leveled its axis against the sun. All day the sun shone directly over the equator, and day and night were perfectly divided as the earth held, for a moment, in geometric and temporal balance. Then its axis tilted and the northern hemisphere began to slowly withdraw into darkness.

The day found Robert Peary and the crew of the *Roosevelt* lodged in the ice near the shore of Cape Sheridan, a bleak thrust of land at the edge of the Arctic Ocean. After arriving in northern waters in midsummer, Peary had visited the western coast of Greenland to acquire Eskimos and dogs. He'd hired twenty-two Eskimo men and their families—seventeen women and ten children—whom he brought aboard the *Roosevelt* to join the nineteen white men and one black man already living on the ship. To this human manifest he'd added a canine cargo of nearly 250 sledge dogs. Sailing low, the *Roosevelt* had threaded through the icebergs of Smith Sound and the thick ice of Kane Basin, arriving at Cape Sheridan in early September. The ship and its crew would pass the winter night here in the clench of the ice pack, hunkered in their small creaking vessel at the edge of the known earth, waiting for the sun to return.

For now, in the waning light of late September, the ship was a hive of industry. The crew lowered and dried the sails and drained the boiler. The Eskimo women set fox traps near the ship and sewed clothes that would be needed for the expedition, using sinews of deer muscle to thread together bearskin trousers, foxskin jackets, fawnskin shirts, and sealskin boots. Hunting parties of Eskimo men forayed into the neighboring wilderness to pursue musk oxen and other meats and pelts. A few days before the equinox, the first sledging party had left the ship and headed west across Cape Sheridan to begin caching supplies for the expedition ahead.

Even in these busy hours, Peary surely found moments to ponder the whereabouts of Frederick Cook. He had reason to be optimistic that Cook, despite a year's lead, had not beaten him to the Pole; or if he had beaten him, had not lived to tell the tale. In early August, when the *Roosevelt* was docked off the coast of Greenland, Peary had met Rudolph Franke, the young German who helped Frederick Cook prepare for his polar expedition and who accompanied the doctor as far as the Arctic Ocean. Franke told Peary that nothing had been heard from Cook in seven months. The chances of anyone surviving seven months on the polar ice, Peary knew, were slim.

In fact, though, Frederick Cook was not dead. He was alive—alive by the skin of his teeth, but alive. Since leaving the Pole in late April, Cook and his two Eskimo companions, Etukishook and Ahwelah, had been lost on the ice. Now, in late September, after five months of wandering, they were about five hundred miles to the southwest of the *Roosevelt*, living in a cave on the desolate coast of Cape Sparbo, on Devon Island, preparing for their own dark days of winter.

The return trek had started out well enough. They made good time and the weather warmed, bringing an end to their eternal shivering. But the warmth also brought new challenges. The polar ice began to break up as the men traveled south, forcing them to skirt wide patches, or leads, of open water. To make matters worse, a thick mist dropped over the men for three weeks, hampering Cook's ability to gauge their location by sextant readings. Only after the air cleared did Cook discover the awful truth: the ice pack had been drifting not to the east, as he'd surmised, but to the west and south. They were fifty miles west of Axel Heiberg Island—fifty miles of broken ice and water—and hundreds of miles from their caches of food and their intended route home.

Cook reasoned that the best course under the circumstances was to follow the shore line south to Lancaster Sound, where they might spot a whaling ship working the northern waters. Their empty stomachs were temporarily filled when Ahwelah managed to shoot and kill a polar bear, but soon they were starving again. In early July, they ran into a lead too wide to circumvent. They had no choice but to abandon the dogs on the ice and squeeze into their small canvas boat. This was, wrote Cook, "the saddest incident of a long run of trouble." As the boat floated away, the dogs "howled like crying children; we still heard them when five miles off shore."

Now followed a fantastic sequence of calamities. During a fierce summer storm, the men took refuge on the hulk of a fast-moving iceberg, riding out the storm by clinging to its slippery edges. The berg made a fine vessel until it suddenly changed course and drifted into the open thrashing sea. By the time the storm died down the next day, the berg had transported the men back to almost exactly the spot they had been two weeks earlier.

They reached Cape Sparbo in early September. They were three hundred miles from the Eskimos' home at Annoatok. In good health and decent weather, they might have made the trek in a month. But with the sun failing and hunger gnawing, they could go no farther. They made camp among the ruins of an old Eskimo settlement, where they discovered a cavelike hollow enclosed by a dilapidated roof. They converted this into a winter home. The human skull they happened upon while fixing up the cave robbed the place of some of its domestic charm, but they were in no position to complain.

Once their den was fit for occupancy, they turned their attention to food. They were nearly out of bullets (Cook held on to four final cartridges, but he was saving these for an unspecified "last emergency") so they improvised with the supplies at hand. They fashioned bows and arrows from pieces of hickory and animal sinews; the blades of a pocket knife became arrow tips. Thus armed, they went hunting. Walrus, hare, and ptarmigan (a bird of the grouse family) were plentiful, and the men laid in as much as they could kill before the sun set for winter.

"The storm of the descending sun whipped the seas into white fury and brushed the lands with icy clouds," wrote Cook. "With the descent of raw sun, nature again set its seal of gloom on Arctic life."

The Unforgettable Season

Twenty-two hundred miles to the south of Frederick Cook's winter camp, the New York Giants were engaged in their own life-and-death struggle on that first day of autumn. Whatever its cosmological or solar significance in other places on earth, the equinox that year had special meaning in New York City—and in Chicago, and Pittsburgh, and more generally across the wider universe of baseball—for reasons having nothing to do with the position of the polar axis. September 23 happened to be the date of a meeting between the New York Giants and the Chicago Cubs. And in the dwindling weeks of the 1908 baseball season, every day the Giants played the Cubs was a special day.

The National League was in the throes that September of a remarkable three-way battle for the pennant. The Giants, the Cubs, and the Pittsburgh Pirates had been neck and neck for weeks as game after game went down to the wire. In the subtitle of his book *The Unforgettable Season*, G. H. Fleming described the 1908 National League runoff as "the greatest race of all time." It certainly seemed that way to the people living through it. Nationally, a record eight million Americans would attend a major league baseball game in 1908. Almost a million of these fans would pile into the Polo Grounds alone, setting a single-stadium attendance figure that would go unsurpassed until 1920, the year Babe Ruth arrived at Yankee Stadium across the Harlem River.

Baseball was a critical piece of American popular culture in the early twentieth century. To civic leaders and progressives like Jane Addams, its rising popularity was applauded. They appreciated the sport's capacity to bind Americans together in good, clean, rule-abiding fun. Baseball was "patriotically patriotic," is how one Chicago civic leader put it that year. But there was another side to baseball these types of encomiums tended to ignore; an element of the national pastime that had more in common with riots, frankly, than sports. This element was very much in evidence in 1908, a season of tremendous excitement and considerable anarchy.

It is tempting to look to the many remarkable events occurring beyond the Polo Grounds that year—the Wright Brothers' flights, the automobile race around the world, Peary's journey to the Pole, the presidential election—as contributors to the feverish intensity of baseball fans. But it's entirely possible the true baseball "fanatics" had no idea these other events were occurring, so absorbed were they in the triumphs and catastrophes of their teams.

Baseball's popularity that year probably owed something, at least in the early months of the season, to the high unemployment rate in the country; even a jobless man might scrape together a few nickels to kill an afternoon at the ballpark. As the season progressed, though, and as mere diversion became rabid obsession, it had ever more to do with the stupendous level of play and the drama of the pennant race.

Each of the three National League teams vying for the pennant had a legitimate claim on all-time greatness. In the case of the Pittsburgh Pirates, that claim rested squarely on the wide shoulders of future Hall-of-Famer Honus Wagner, probably the best shortstop who ever played the game. The Chicago Cubs were blessed by their own Hall of Fame–bound infield: shortstop Joe Tinker, second baseman Johnny Evers, and first baseman/manager Frank Chance. The devastating double-play threat represented by these three men was immortalized in a poem by New York writer Franklin P. Adams:

> *These are the saddest of possible words—*
> *Tinker to Evers to Chance.*
> *Trio of Bear Cubs and fleeter than birds—*
> *Tinker to Evers to Chance.*
> *Thoughtlessly pricking our gonfalon bubble,*
> *Making a Giant hit into a double,*
> *Words that are weighty with nothing but trouble:*
> *Tinker to Evers to Chance.*

Another future Hall-of-Famer on the Cubs that year was Mordecai Peter Centennial Brown, better known to baseball fans as "Three-Fingered" Brown, on account of a childhood accident in which two fingers of his pitching hand had been lopped off by a corn thresher. Brown used the stub of his missing index finger to release the ball with a confounding curve, "the most devastating I ever saw," in the words of Ty Cobb (who had recently started his own astonishing career with the Detroit Tigers). With its impressive roster, the Cubs had won the National League pennant in both 1906 and 1907.

But no team was more consistently superior, or more persistently colorful, than the New York Giants. The tone of the Giants was set by the team's manager, John McGraw, a salty-mouthed bulldog of a man. "Muggsy," or "Little Napoleon" as he was sometimes called (he measured just five and a half feet), had been an extraordinary ball player as a young man; now, at thirty-five, he'd evolved into a brilliant manager. He had assembled a championship team of hard-living roustabouts as scrappy and foul-tempered as himself, including second baseman "Laughing Larry" Doyle, pitcher Joe "Iron Man" McGinnity, and captain "Turkey Mike" Donlin.

The star of the Giants was a man who for all appearances did not belong on the team. Christy Mathewson, better known to baseball fans as "Matty" or, for more obscure reasons, as "Big Six" (perhaps because he was over six feet tall), was one of the first true superheroes of baseball. Handsome, clean-cut, morally scrupulous, college educated, well-read (his tastes included Victor Hugo and William James) and, most remarkable of all, a genuinely nice guy, he did more than any man of his time to bring the wholesome middle class to baseball. Of course, none of his personal attributes would have mattered much had Christy Mathewson not also possessed one of the finest arms to ever pitch. Since joining the Giants in 1900, he had repeatedly set records of achievement, including one in 1905 for throwing three consecutive shutouts, a feat that never has been repeated since.

In 1908, at twenty-eight, Matty was having the finest season of his career. Before the season was done, he would win thirty-seven of the forty-eight games he pitched, including twelve shutouts. This at a time when pitchers usually pitched *entire* games; Mathewson pitched forty-four of his forty-eight from beginning to end.

Given all the fabled players on the field that day, it's ironic that the man whose name would forever be associated with September 23—and with that entire 1908 season—was the least well-known of the lot. Nineteen-year-old Fred Merkle was a promising new recruit from Wisconsin who had joined the team at the end of the previous season. He had played well in the April opener at the Polo Grounds but had spent most of his time warming the bench since then. It was only by chance that he was in the game at all on this Wednesday afternoon. First baseman Fred Tenney had come down with a case of lumbago, making him the latest casualty to join the list of "the cripples," as the press had taken to calling the Giants' roster of injured players. McGraw turned to Merkle for a fresh pair of legs. This would be Merkle's first-ever start for the Giants. His lucky break.

The game began at 2 P.M. The Polo Grounds was packed with edgy fans. A few days earlier, after the Giants beat the Pirates, the *New York Tribune* had declared it "practically a foregone conclusion" that the Giants would take the pennant, but that assessment was looking less and less foregone every day. As of game time on September 23, the Giants' 87 wins and 50 losses put them just a few percentage points ahead of the Cubs' 90 wins and 53 losses. The Pirates were just a few points behind the Cubs. Every game was potentially decisive.

McGraw assigned Christy Mathewson to pitch for the Giants. Matty was his usual superb self. Young Fred Merkle, too, was terrific through the first eight innings. In nearly every inning he fielded an out, including one spectacular catch of a low line drive.

Like so many games before it that year, this one was a nail biter, another "bang up, all a-quiver game," as the *New York Times* sportswriter W. W. Aulick put it in the next day's edition. The score was tied 1–1 as the Giants came to bat in the bottom of the ninth inning. First up was Cy Seymour, who grounded out. Next, Art Devlin got a single. Harry "Moose" McCormick hit a single, too, but Devlin was forced out at second. With two outs and a man on first, Fred Merkle came to the plate. The pressure on the young ball player was tremendous. If he struck out, the game remained a tie and went to extra innings.

Merkle came through in the clutch: he hit a long single to center field. The crowd cheered wildly as Merkle ran to first, McCormick advanced to third, "and everybody in the inclosure slaps everybody else and nobody minds," wrote W. W. Aulick. "Perfect ladies are screaming like a batch of Coney barkers."

Now there were two outs and two men on base. Al Bridwell, the Giants' shortstop, was up. On the first pitch, he slammed a line drive into the grass of center field. This was all the Giants needed to win the game. Bridwell ran for first, McCormick ran home to post the winning run, and Fred Merkle—here is where it gets complicated.

Overcome by confusion or by force of habit, or perhaps by the heady pride he took in having made a success of his first-ever start for the New York Giants, Fred Merkle apparently ran halfway to second base, then turned and dashed for the Giants' clubhouse. Many of the details would later come up for dispute—how far Merkle ran toward second, where

exactly he went—but by the time the dust settled, the clear consensus was
that Fred Merkle never made it to second base. In another game, against
another team, it's possible nobody would have noticed or cared that
Merkle did not run to second. Touching second base was, under the circumstance, a formality, rather like a golfer putting a ball an inch from the
rim of the hole—a shoo-in, a gimme. And in those more organic and fluid
days of baseball, the finer points of protocol were often bypassed in
favor of more pressing interests, like getting to the clubhouse before
twenty thousand fans could scramble onto the field and maul every man
in a uniform. Indeed, even as Merkle was running for the clubhouse, several Chicago Cubs were doing the same.

Unfortunately for Merkle and the Giants, some Cubs stayed behind.
More unfortunate still, a few of them noticed Merkle's oversight. And in
that do-or-die season, they were in no mood to let it pass. Merkle may
have had custom and expediency on his side, but the Chicago Cubs had
the rules on theirs, and the rules were clear: if the ball got to second base
before Merkle got there, he was out. And if Merkle were out,
McCormick's run did not count and the game remained a tie.

Now began a frantic effort by Chicago to get the ball to second base
before the Giants could realize what was happening and retrieve Merkle.
Chicago center fielder Solly Hofman threw the ball to Johnny Evers at
second base, but in the melee already sweeping across the field the throw
missed Evers. At this point, Giants pitcher Joe McGinnity, who had
been coaching third base, grasped the situation. Not nicknamed "Iron
Man" for nothing, he leaped into action, forcefully grabbing the loose
ball (in one version, yanking it from Johnny Evers's glove). As Frank
Chance and several other Cub players tried to wrest it back from him,
McGinnity broke free of the Cubs and hurled the ball into the stands
behind third base. Joe Tinker ran to the stands to retrieve it. A few
moments later, somehow—the details are murky—the ball reappeared,
now snug in Evers's glove. More accurately, *some* ball appeared in Evers's
glove; that it was the *same* ball McGinnity threw into the stands seems
unlikely. In any case, Evers placed his foot on second base and began
shouting that Merkle was out.

Neither the home plate umpire, Hank O'Day, nor the field umpire,
Bob Emslie, had seen Merkle touch second base; then again, neither had
seen him *not* touch second base. As Cubs players yelled in one ear that
Merkle was out, and Giants players yelled in the other that Merkle was

safe, Hank O'Day seemed to waffle. At last, he made his call: Fred Merkle was out. McCormick's run did not count, therefore, and the game remained tied at 1–1.

Technically speaking, if the game were tied, it should have gone on into extra innings then and there; practically speaking, though, resuming the game was out of the question. The field was flooded with bewildered but irate fans. Many aimed their rage at the Cubs players. Others went directly for Hank O'Day. "[A]lthough most did not know what it was all about," reported the *Herald*, "everyone evidently recognized a good opportunity to get a shot at the umpire."

The following day, the New York sports press was divided in its response to the game. The *New York Tribune* cast its allegiance with the Giants— "Never was a victory more cleanly won"—while the *Times* lit into Merkle for committing an act of "censurable stupidity." The *Herald* advised Merkle to "gather the idea into his noodle that baseball custom does not

The Polo Grounds on September 23, after the Merkle debacle.
Coogan's Bluff rises in the background.

permit a runner to take a shower and some light lunch in the clubhouse on the way to second," but allowed that what Merkle had done was, after all, fairly routine. Sam Crane of the *New York Evening Journal* declared his conviction that Merkle *had* touched second base, and that all the rest was only a devious ploy by the Cubs to steal victory from the Giants.

The usually measured Christy Mathewson voiced the outrage of many New York fans in the *Evening Mail*. "If this game goes to Chicago by any trick of argument," he was quoted as declaring, "you can take it from me that if we lose the pennant thereby I will never play professional baseball again."

On one thing, at least, all the newspapers could agree: Merkle's blunder had turned a merely thrilling season into a magnificently strange one. "The National League race is now a lulu—" declared the *New York American,* "the luluest kind of a lulu what is."

"It had not seemed possible that there could be any increase of interest in baseball," the *New York Times* observed in an editorial, "but just now it is the subject that seems to claim most of the attention the multitude of our fellow-citizens can spare from business and domestic affairs."

"No similar situation," the *Herald* stated (slipping into the pluperfect as if the Merkle game were already of the fabled past) "had ever been seen in the history of baseball."

The Experimental Room

The traumas of baseball that year reached beyond New York, beyond Chicago and Pittsburgh, beyond the entire National League, in fact. Far away from New York, in another principality of baseball, the American League Detroit Tigers were in the throes of their own gut-wrenching struggle that fall. Led by twenty-one-year-old wunderkind Tyrus Raymond Cobb, the Tigers were doing their best to fend off that other superb Chicago team, the White Sox, to win the American League pennant.

The Tigers were a rising team in a rising city. Only recently, Detroit had been more of a bustling backwater than an industrial powerhouse. Now its population of more than four hundred thousand made it the ninth-largest city in the country (according to the census taken two years later). And its position as the automotive capital of America, coming into ascendancy at the very moment the automobile was about to transform the country, made it one of the fastest growing. On October 1, 1908, as the Tigers were plowing their way through the middle of an astonishing ten-game winning streak in the city's Bennett Park stadium, that position was secured by the debut of a new automobile from the Ford Motor Company.

The new automobile did not look like a machine of any great consequence. On the contrary, it was a prosaic-looking affair, boxy and top-heavy, "[u]ncompromisingly erect, unquestionably ugly, funereally drab," as the automobile writer Floyd Clymer would later describe it. The interior only confirmed the bumpkinish first impression. The hard-sprung, church-pewish seats made no concession to elegance or comfort. The location of the steering wheel, on the left-hand side of the vehicle, was disconcerting at a time when most steering wheels were located, European-style, on the right. The advertised top speed of forty miles per hour, when sleek cars like the Thomas Flyer or the Locomobile could easily achieve sixty, was an insult to modern impulses.

And then there was that name: that solitary letter, and not an especially compelling letter at that. Not a sinuous S or an inquisitive Q; not an off-kilter R or elliptical O or assertive X. Just a symmetrical, perfectly balanced, perfectly insipid T.

Henry Ford had already experienced a good deal of success when he began producing the Model T in the fall of 1908. At forty-five, he'd been in the automobile business a dozen years, since building his first horseless carriage in a brick shed behind his Detroit home in 1896. It was not auspicious that Ford, whose very name would someday be synonymous with industrial efficiency, failed to notice that the door of the shed was too narrow to get his machine out until he'd finished building it. And yet this error of planning gave him an opportunity to demonstrate his genius for finding simple solutions to complex problems: he picked up an ax and bashed a wide hole through the brick wall.

Thus, in early June of 1896, at about four o'clock in the morning, Henry Ford drove his four-horsepower "quadricycle" out into the dark streets of Detroit, rousing the sleeping houses of the city with the loud sputtering homemade internal combustion engine. He was launched.

Ford's early attempts to link his mechanical genius to financial profit were a mixed success. Despite notable innovations, his first two automobile companies floundered. It was his third try, Ford Motor Company,

founded in 1903, that turned out to be the charm. By 1908, he was manufacturing about 115 cars a week, a production rate second only to Buick,
and employing nearly two thousand men at a plant on Piquette Avenue
in Detroit. From the Model A, which he'd begun producing in 1903, he'd
skipped his way through the alphabet, improving the product with each
new effort. He'd earned a reputation in Detroit as an innovative if somewhat eccentric force in the automobile industry, and his products had
earned a reputation for reliability, durability, and affordability.

But Ford was not content. Everything he had done was a warm-up to
what he really hoped to accomplish. "I will build a motor car for the
great multitude," is how he stated his great ambition.

> It will be large enough for the family but small enough for the individual
> to run and care for. It will be constructed of the best material, by the best
> men to be hired, after the simplest designs that modern engineering can
> devise. But it will be so low in price that no man making a good salary
> will be unable to own one—and enjoy with his family the blessing of the
> hours of pleasure in God's great open spaces.

Here was Ford's holy grail: an automobile that would be affordable and
useful to Americans heretofore excluded from the automobile frenzy; the
sort of automobile a hardworking middle-class farmer in Dearborn,
Michigan—a man like his father, for instance—could purchase without
hardship or regret; a model that would cost under a thousand dollars and
yet perform as well as—no, better than—an automobile costing two
thousand dollars; a machine that would burst wide the range of possible
buyers, put a car in every driveway and a family in every car, and make
Henry Ford one of the richest men in the world.

Actually, getting rich was never what drove Henry Ford. He was a man
of simple tastes who disdained ostentation. The vices of the idle rich—
gluttony, tobacco, alcohol, divorce—held no appeal to him (a trait he
shared with fellow inventors like Thomas Edison and the Wright brothers). Farming had never appealed to him, either, but he prided himself on
being a farmer's son and felt only contempt for the conspicuous consumption of the rich. Consumption by the working class and middle class,
though—now that was something else entirely. Particularly when the consumption involved the purchase of a Ford automobile.

Ford advanced an all-American political philosophy of democratic consumerism. Whereas socialists wished to level the rich, Ford intended to give the masses the means to live as if they *were* rich. Ford would never have compared himself to the crass impresarios of Coney Island, but he was working along similar lines, granting unwealthy Americans the pleasures of wealth at an affordable price. In that political season of the fall of 1908, this was a message far more potent than anything William Jennings Bryan, William Howard Taft, or even the Socialist candidate, Eugene Debs, were offering in their respective stump speeches.

Of course, it was one thing to promise an inexpensive but superior automobile. It was quite another to deliver it. There were already cheap motorcars on the market, but most were shoddy products, for the simple reason that good cars were expensive to manufacture. To succeed, Ford had to take two seemingly contradictory aims and make them compatible.

In early 1907, Ford had escorted one of his favorite managers, Charles Sorensen, to the northern end of the third floor of the Piquette Avenue plant. "Charlie, I'd like to have a room finished off right here in this space. Put up a wall with a door big enough to run a car in and out," Ford instructed Sorensen (having apparently learned his lesson about cars and doors). "We're going to start a completely new job."

The locked room became known as the "experimental room." Just fifteen by twenty feet, it was a cramped lair. Only half a dozen men were allowed access, including a master draftsman named Joseph Galamb, a few assistants, Sorensen, and Ford. In an intensely collaborative process that historian Douglas Brinkley has compared to the Manhattan Project, the men brainstormed for endless hours amid car parts, power tools, and chalkboards. As Ford presided over the sessions from a large rocking chair, ideas flew around the room, captured by chalk on slate. Sketches of parts were tooled into physical prototypes so that Ford could see and feel them physically. He had little use for blueprints. What he did have, once he got his hands on a thing, was a genius for understanding how it worked, and how to make it work better.

In addition to his mechanical genius, Ford brought to that room a number of convictions. Chief among these was his insistence that the solution to producing a high-performance car lay not in boosting horsepower, which added to the expense both of building and maintaining the machine, but in lowering weight. Lightness *was* power. His twenty horsepower machine would (as Ford put it in 1908) "go anywhere a car of double the horsepower will" but would not be a "wrecker of tires," as heavy cars so often were. A light car also required less gas and oil to operate, a great advantage at a time when fuel was still purchased in cans from hardware stores.

Ford's decision to use vanadium steel in the Model T was among the most critical he made. Steel alloyed with the element vanadium had a tensile strength nearly three times that of regular steel. Vanadium steel was not new, but Ford was the first to appreciate its applications to a mass market automobile. The high-strength alloy would allow the car to be lighter—a Model T weighed in at twelve hundred pounds, about half the weight of the Thomas Flyer—and would allow Ford to purchase less steel per vehicle than most cars consumed. This was the very essence of Ford's genius. He figured out how to cut costs without cutting corners.

Every aspect of the car was considered with an eye to simplicity, inside and out. The simpler a piece of machinery, Ford understood, the lower the cost of manufacturing it, and the easier and cheaper the task of maintaining it. Equipped with a manual and a few basic tools, a Model T owner would be able to carry out most repairs himself. The new car's planetary transmission would be smoother and longer lasting than any that had ever been designed. The magneto, a small magnetized generator that provided a steady flash of voltage to ignite the automobile's fuel, would be more dependable.

More obvious were the bodily changes on the exterior, those characteristics that made it appear so ungainly at first sight. The Model T was designed to ride high off the ground to give it plenty of clearance over America's infamously bumpy roadways, while the car's three-point suspension system allowed it to handle the roads without tossing its occupants into a roadside ditch.

Ford's concessions to the contemporary realities of American roads notwithstanding, he also foresaw a day when the ditch at the side of the road would be less of a concern to motorists than oncoming traffic. He therefore moved the steering wheel over to the left side of the vehicle in order to improve the driver's view of approaching traffic. He designed cars as if he already knew for a certainty that within seven years 2.5 million automobiles would be on the roads in America—and that half of these would be Model Ts. In 1908, even as he was producing the first prototypes on Piquette Avenue, he boldly began constructing a new plant on the site of an old horse track in the Detroit suburb of Highland Park. This new plant, when completed in 1910, would dwarf every automobile factory that had come before it.

Not everyone shared Ford's faith. Many believed that he was recklessly overextending the company. "Some of the stockholders were seriously alarmed when our production reached one hundred cars a day," he later wrote in his memoir. "They wanted to do something to stop me from ruining the company, and when I replied to the effect that I hoped before long to make a thousand a day, they were inexpressibly shocked." Over the next few years, the stockholders who held their breath—and held on to their shares—would have the pleasure of being shocked all over again.

Henry Ford was out of town, hundreds of miles from Detroit, when the Model T made its public debut. He'd gone north to the Upper Peninsula of Michigan on a hunting trip with several men, driving Model T serial number 00001, the first production model. Absenting himself from Detroit at the most important moment in his company's history might have been a sign of confidence or evidence of his talent for delegating responsibility. Possibly it was a case of jitters, like a playwright fleeing the theater on his play's opening night. Ford had sunk everything he had into this new automobile. His future, and the future of Ford Motor Company, depended on its success.

The company launched a national ad campaign starting that first day of October. Full-page ads appeared in magazines like the *Saturday Evening Post* and *Harper's Weekly*. The style of the ads, like the style of the product itself, was plain but carefully considered, pitched with a perfect ear for the aspirations of its intended audience. *"We have no high-sounding names with which to charm sales. It's the same old name, 'plain as any name could be;' it's just 'FORD.'"*

For a stripped-down, "unheard of" price of $850 (an extra hundred would get the buyer such amenities as a windshield, speedometer, and headlights), the ads promised "a 4-cylinder, 20 h.p., five passenger family car—powerful, speedy and enduring."

"Your guarantee that this car is all we claim—and our claims are broad—is in the reputation of Henry Ford." Seldom would a man and a mass-produced machine be so inseparable. The machine was an expression of its maker's worldview, an extension of his personality. It was Henry Ford to a T.

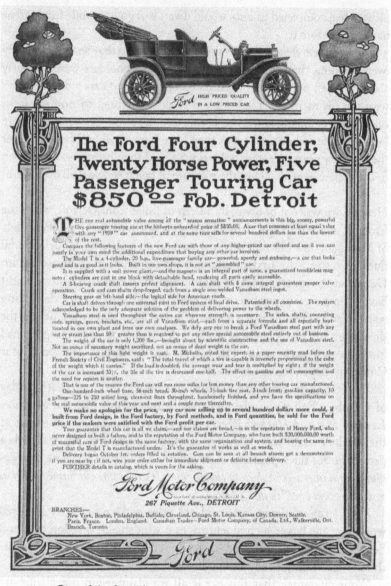

One of the first ads to pitch the Model T to the public,
this appeared in *Harper's Weekly* on October 3, 1908.

The ads worked. Orders began pouring in at once, more than the factory on Piquette Avenue could possibly fill. Not even Ford could have predicted the success of what he had created. Certainly he could not have anticipated the effect the Model T would have for years to come on how Americans lived, worked, ate—on the landscape of the country and the air they breathed—on nearly every aspect of American life. Douglas Brinkley's comparison of those sessions in the experimental room on Piquette Avenue to the Manhattan Project was apt. Not until Robert Oppenheimer and his fellow scientists met in secret to develop the A-bomb would Americans produce a device of such transformative force. The effects of the Model T worked more slowly than those of the A-bomb, but they were no less explosive.

The Great Game

On the evening of October 2, New Yorkers observed "peculiar and weird" flashes of light emanating from the top of the nearly completed Metropolitan Life Building on Madison Square. The lights appeared to be some kind of semaphore, sending coded messages into the night. Sure enough, they were soon answered by similar flashes from the vicinity of the Eighth Regiment Armory far uptown and the Seventy-first Regiment Armory at Park and Thirty-fourth. Further investigation revealed the lights to be the work of the Army Signal Corps. "Members of the corps were engaged in studying a private code of signals," reported the *New York Herald*, "to be used if this country ever again goes to war with a foreign power."

The following evening, Madison Square Garden kicked off its second annual Electrical Show. Displays included such futuristic contrivances as automated cow-milking machines, automated egg-hatching machines, and automated house-cleaning machines. Thomas Edison delivered the keynote address that evening without ever appearing in person. His recorded voice spoke to the audiences from a phonograph on the stage.

Among those attending the exhibition at the Garden were members of the Esperanto Association of North America. The Esperantists, as they called themselves, looked to the exhibition as an opportunity to broaden the public's interest in their "International Auxiliary Language." A linguistic salad of Latin roots and bits of English, Russian, and German, Esperanto was an invented language meant to give humanity a tool to communicate across borders and cultures. If Esperanto were adopted worldwide, harmony and peace would ensue. That, anyway, was the general idea.

In a city undergoing massive immigration, and which had just completed the two tallest Towers of Babel in the world—the Singer and the Metropolitan Life buildings—the idea of returning to a prelapsarian state of linguistic harmony had a nice ring to it, even if the language itself, a sort of guttural pig Latin, did not. As for what this had to do with mechanical cow-milkers and egg-hatcheries, an Esperantist was on hand to explain: "Electricity is the quickest and most modern force of its kind. Esperanto is the quickest and most modern language. You see why we are here."

The most eye-catching display of electricity in New York that first weekend of October was neither the flashing searchlight of the Metropolitan tower nor the automated cow-milker in Madison Square Garden. It was the "electrical baseball boards" at the Polo Grounds.

On Sunday, thirty-five hundred fans rode out to the ballpark to watch the Chicago Cubs play the Pittsburgh Pirates. They could not watch the actual game, of course; that was occurring a thousand miles away in Chicago. What they came to see, instead, was a real-time simulacrum played out on two large boards, each marked with a baseball diamond and festooned with lightbulbs. The boards stood near home plate, facing the crowd in the grandstands. Some lights on the board indicated where players stood on base or in the field; others showed where the ball had been hit. As telegrams flowed in steadily from Chicago, the board's operators translated the words into flickers of light. The electric light board was a far cry from a real game, but it was as close to watching a live play-by-play broadcast as anyone could get in 1908. Among those in the stands watching the boards were a few of the Giants players, including Christy Mathewson. Like the rest of the crowd, they rooted for the Cubs to win. Their hopes, which had seemed assured just weeks earlier, now depended on a Chicago victory.

The days since Merkle's blunder had not been kind to the Giants. After several losses, New York had fallen from first place into third. Pittsburgh was now in first, half a game ahead of Chicago. If Pittsburgh won today, the Giants' hopes were dashed. But if Chicago won, the Cubs moved into first and the Giants remained alive.

So as the game flickered over the boards, Giants fans, all turned into rabid Cubs fans for the day, cheered Tinker's game-opening double in the fifth, Chance's amazing catch in the eighth, and every strikeout thrown by Three-Fingered Brown. When it was over, Chicago had won 5–2, the Giants were still in the pennant race, and New York fans could get back to hating the Cubs.

The next day, the National League Board of Directors met in Cincinnati to deliberate about the Merkle game. The president of the National League, Harry Pulliam, had earlier ruled the game a tie, but both the Giants and Cubs organizations had issued formal protests. The final decision rested with the board. In the meantime, the teams and their fans could only wait. "A hush fell upon 80,000,000 people," wrote a reporter in the *New York Evening Mail,* perhaps overstating the case just a little. "The wheels of industry had ceased their whirr . . . husbands halted on their homeward ways, wives let the dinner grow cold . . . in a stilled and silent wait for the decision on that tie game." That same Monday, the Giants began a three-game series against the Boston Doves. The Doves were a mediocre team, but the pressure on the Giants was extreme. They could not afford to lose a single game or their hopes for the pennant would be dashed.

On Tuesday, October 6, the Detroit Tigers beat the Chicago White Sox 7–0 to win the American League pennant and fill one half of the slate for the upcoming World Series. Chicago's hopes for baseball glory now rested entirely on their Cubs—and on the board's decision.

It came that evening. The board ruled the Merkle game a tie. Whatever outrage this decision might incite, the board implied, should not be directed at itself but at the man who caused the debacle in the first place. "The game should have been won for the New York club had it not been for the reckless, careless, and inexcusable blunder of one of its players—Merkle," the board stated in its decision. The board ordered the Cubs and Giants to replay the game to resolve the tie, setting Thursday, October 8, as the date of the final match. Like the September 23 game, this one would take place at the Polo Grounds.

The response to the board's ruling among the Giants and their fans was bitter outrage. John McGraw told reporters that anything short of a Giants victory was "highway robbery." Some New Yorkers suggested that the Giants should boycott the rematch on principle. Others swore that if the Giants did play and went on to lose the pennant as a result, they would never set foot in the Polo Grounds again.

No one could have been more devastated by the board's decision, of course, than Fred Merkle. His days had been consumed by humiliation and contrition since September 23. No baseball article was complete without reference to his blunder, his "bonehead" move—"Merkle's boner," as it came to be known. The piling on was ruthless and Merkle took it hard. Rumors floated around the city that he was losing weight, that he had committed suicide. John McGraw, to his credit, continued to assure the young player that he'd done nothing wrong. When Merkle offered his resignation, McGraw refused it.

Secretly, Merkle must have prayed that the Giants would lose enough games in what remained of the season to make the September 23 outcome, and his blunder, immaterial to the team's pennant hopes. If so, his prayers went unanswered. The Giants won all three games against the Doves. After their last regular season game on Wednesday, October 7, their record was ninety-eight wins and fifty-five losses—exactly the final record of the Chicago Cubs. The October 8 replay, then, would decide the outcome not only of one game; it would decide the entire season. The winner would take the pennant.

Giants fans may have been upset by the board's ruling, but even the most ardent of them—especially the most ardent—had to appreciate the sublime drama it had created. "It is a bombastic and fitting climax to a season of unparalleled thrills," observed the *New York Herald*. "Never before has the race been so close. Never has it been necessary to play off the tie of six months' baseball in a single gigantic battle." The game would be, the *Chicago Tribune* agreed, "[t]he biggest day that baseball ever saw." Even the relatively sedate *New York Times* reached for superlatives equal to the occasion: "Perhaps never in the history of a great city, since the days of Rome and its arena contests, has a people been pitched to such a key of excitement." *Harper's Weekly* called it simply "The Greatest Ball Game Ever Played."

On Wednesday afternoon, the Chicago Cubs boarded the Twentieth Century Limited, waved farewell to the many well-wishers who had come to see them off, and began their overnight journey to New York. They were still speeding through the dark hinterlands when young men and boys began showing up at the Polo Grounds in upper Manhattan. The stadium's management had posted extra watchmen for this eventuality. The watchmen were kept busy throughout the night, brushing would-be intruders back from the fences or hunting them down under seats, where some of the smaller boys were crouched in hiding, prepared to wait until the following afternoon.

By dawn, the aspiring entrants had multiplied exponentially; by late morning, a crowd like none ever seen at the Polo Grounds was pressed against the gate. And still more people came in waves behind them, tumbling off elevated trains so crowded some passengers rode outside on the roofs of the cars.

When the gate opened at 11 A.M., four hours before game time, people rushed into the Polo Grounds like fish pouring through a breached dike. Some came with tickets, others with nothing more than gall and quick feet. The risk was small for scofflaws. Police Commissioner Bingham had dispatched three hundred New York City policemen to assist the regular Polo Grounds security force on this most remarkable of game days, but *three thousand* would hardly have been sufficient. At 12:45, two and a quarter hours before game time, the stadium was already dangerously overcrowded with paying and nonpaying customers. The gates were shut to prevent more people from entering. Newly arrived fans found their way blocked by cold metal and pitiless policemen.

"But, Officer," some would plead, "we have reserved-seat tickets. We have paid money for them, and the seats are waiting for us to occupy them. Don't we get in?"

"Certainly youse don't," the policeman would respond. "Youse can't get in and that's all!"

Away from the gates, "speculators" ran a brisk trade in selling tickets for nonexistent seats to unsuspecting fans. When it became clear to those who bought them that they'd been duped, they returned for vengeance. Fights flared up. "But the speculators kept selling—and fighting," reported the *Evening Telegram*. "They were getting such prices that they could afford to throw in a little fight with each ticket." Meanwhile, pickpockets slithered through the crowd, harvesting wallets and jewels.

By 2 P.M., an hour before the game, as many as a quarter of a million people had traveled to the Polo Grounds. Thirty-five thousand were already inside, but determined and desperate fans continued to besiege and assail the stadium. They scaled fences, dug under fences, tore down fences, charged fences in flying wedges, and, in one case, burned down a fence. Polo Grounds officials turned hoses on the vandals to keep them at bay but with limited success. Hundreds of panting, drenched boys and young men slipped into the stadium and vanished into the crowd.

Outside the stadium, thousands of men and women packed the edge of Coogan's Bluff, the high escarpment that rose behind the grandstand. Thousands more climbed onto rooftops, treetops, smokestacks, chimneys, telegraph poles—any object that rose a few feet and provided an angle from which to peek into the stadium. "From the grandstand the sky was one mass of human beings," reported the *Herald*, "and, although the day has not yet arrived when airships are ordinary, the higher altitudes about the Polo Grounds were very thickly populated yesterday."

The situation was mirthful but perilous. An off-duty fireman fell to his death from a pillar on the El tracks at Eighth Avenue and 159th Street. "Only the vigorous use of clubs by the police cleared a small circle around the dead man and kept others from climbing the pillar," reported the *Evening Telegram*. Others were injured in falls and brawls. When ambulances arrived at the gate to cart away the casualties, people outside offered rich bribes to the drivers to ferry them into the Polo Grounds.

The players, meanwhile, began to stroll onto the field. They were accustomed to playing before big crowds, but none like this—there had never been one like this. The fans pressed in close, the tens of thousands in the stands, thousands more standing (as permitted in those days) at the perimeter of the outfield, almost breathing down the necks of the players. As the first Giants appeared from the clubhouse, the crowd cheered loudly. Then came Merkle, his head down—"a melancholy figure," the *New York Sun* described him—and the crowd quieted as if a

Outside the Polo Grounds on
October 8, fans climb for
a view of the game inside.

dark chill had passed over the Polo Grounds. Merkle crossed the field and began to toss the ball with his teammates. "Nobody had the heart to jeer him," recorded the *Sun*. "But all the same—"

More than half the Giants were lame from a variety of orthopedic injuries, but they had roused themselves for this last battle. "I have seen all the boys," McGraw told the press before the game. "Everyone said he would play as he has never played before. If we lose we will die fighting and with our boots on, but we're going to win."

The volume in the stadium rose again when Christy Mathewson, looking princely in a white duster, appeared from the clubhouse. As Chicago took the field, the cheers turned to hoots and catcalls and hisses. The Cubs had been on a train all night. They had received death threats before the game and now found themselves pelted with insults and wads of paper and other projectiles by thirty-five thousand people who seemed to wish them the gravest kind of harm. The Polo Grounds, Mordecai Brown later wrote of that day, was "as close to a lunatic asylum as any place I've ever seen."

* * *

Some of the lucky few who made it into the Polo Grounds for the October 8 playoff between the Giants and the Cubs. Their smiles would soon be gone.

It was probably inevitable that the game itself, once commenced, would not live up to the frenzy surrounding it. No game could. Still, this one came close. The most thrilling episodes occurred early. Christy Mathewson pitched beautifully for the Giants in the first inning, shutting the Cubs down. When it was the Giants' turn at bat, they faced the Cubs' Jack "Giant Killer" Pfiester, the same man who had pitched to them in the Merkle game. Almost immediately, Fred Tenney got on base, then Donlin hit a double, sending Tenney home for a run. Pfiester walked another Giant before Chance, taking no chances, pulled him and put in Mordecai Brown. That it should be "Big Six" Mathewson and "Three-Fingered" Brown in this last game—the two dominant pitchers in the league in the greatest of all games—seemed almost fated.

With the dramatis personae now in place, the play carried on. The first inning ended with the Giants up one–nothing. Fred Merkle, back in his role of warming the bench, was seen to smile in the dugout.

By this point, all those millions of New Yorkers who were unable to travel to the Polo Grounds on that Thursday afternoon had found their way to newspaper bulletin boards and anywhere else updates might be available. Traffic came to a halt along Broadway, as Herald Square and Times Square filled with people who could not be induced to move. Down on Wall Street, several of the stock exchanges turned their quotation boards into temporary scoreboards.

In Chicago, a large crowd, including most of the city government, gathered at Orchestra Hall to watch the game on one of the new electric boards. Fresh news was quickly conveyed throughout the city the old-fashioned way, too. "Thousands of girls in skyscrapers by previous arrangement with office boys had bulletins flashed by special signals to them in lofty windows," reported the *Herald.* Meanwhile, "In every city and hamlet in the United States that could boast of telegraph wire, frenzied fandom . . . hung on the ticker . . ."

The telegraph systems weren't the only tickers thrumming that day. According to a report later issued by the New York City Health Department, cardiac deaths spiked in the city. Health officials attributed these deaths to baseball-induced stress.

In the third inning, tickers throughout the land, telegraphic and cardiac alike, began to sputter, and the smile left Merkle's face. Joe Tinker of the Cubs had hit a triple into center field—an easily caught ball that was badly handled. Then Matty's famous control deserted him. He walked a man. The Cubs capitalized with a few more hits and scored a quick succession of runs. Suddenly, Chicago was up 4–1.

Three more innings passed without much drama, but the seventh inning brought the Giants achingly close to the brink of victory. They loaded the bases with no outs. They managed to make another run on a sacrifice fly, but that was all. The score remained 4–2 after the eighth. Their final at-bat in the ninth inning went quickly, mercifully: one-two-three. Just like that, the game, the unforgettable season, was over. The Giants were done for the year.

The Cubs were off to the World Series against the Detroit Tigers, where they would triumph again and post their second world championship in a row (and their last). For the moment, though, they simply had to get to the clubhouse before they could be caught by Giants fans. A few were too slow and were pummeled, including manager Frank Chance, who sustained a hard blow to the neck. Even after the Cubs made it inside the clubhouse, bloodthirsty fans clamored at the door like a lynch mob. Only the efforts of the New York police, who pulled their revolvers to keep the crowd at bay, protected the Cubs from further assault. The Chicago players quickly dressed in street clothes. One by one, they pulled their hats low and slipped into the crowd undetected.

While their foes fled and their fans rioted, the Giants moped and rationalized. "I do not feel badly," said John McGraw. "My team merely lost something it had honestly won three weeks ago." Christy Mathewson lingered in the dressing room long after the game. "I did the best I could," he told reporters when he finally left, "but I guess fate was against me."

Fate did seem to play a hand in that strange season, when the New York Giants lost the pennant to a team led by a man named Chance; when the entire season turned on an impetuous choice made by a nineteen-year-old ballplayer on a day the earth's axis leveled against the sun.

Poor Fred Merkle hardly deserved the scorn, but he got it anyway. He would eventually go on to have a respectable career in baseball, much of it with the Giants, but his name would forever be associated with the game of September 23, 1908. Until the day he died, his nickname would be "Bonehead."

Christy Mathewson would keep playing at the Polo Grounds despite his vow to never return if the Giants lost the pennant. He continued to display outstanding mastery as a pitcher but would never pitch another season as well as he pitched in 1908. As for all those Giants fans who vowed, along with Matty, to quit the Polo Grounds, it's a good bet a few returned anyway, but they would never witness another season like the one just past. Many years later, the Polo Grounds would host another historic moment in baseball, Bobby Thompson's pennant-winning home run in 1951 against the Dodgers—"The Giants win the pennant! The Giants win the pennant!"—but the men who were young in 1908 would be old by then and it would not be the same.

A pall fell over the city after the loss to the Cubs. New Yorkers could find solace, though, in knowing they had just lived through something extraordinary, a moment of baseball that transcended sport and became more like physics, fusing them, as the *New York Tribune* put it, "by a force as mysterious as that which binds the particles in nature together into a compact, united whole that thought and yelled and moved and breathed as one suffering self." This was, perhaps, what Jane Addams and the civic leaders hoped of baseball after all.

The day after the game, the *New York American,* a populist newspaper owned by William Randolph Hearst, detected in the enthusiasm of the last few weeks the possibility of political wakening, and the foundation of a socialistic Utopia of the sort Eugene Debs, Socialist candidate for president, might have approved. Yes, it was painful to witness the Giants' fall from grace, the *American* allowed,

> But we know now that we *can* become excited, energetically, masterfully excited, and as soon as we understand how properly to apply that tremendous dynamite force to the really important things of life, we will get what we ought to have, individually and collectively, and no thieving corporations, no swinish bosses, no bludgeon-bearing election thieves can stand a minute before us.
>
> Merkle's blunder cost New York the pennant . . . But it evoked excitement. No human being in New York yesterday can deny that. And excitement makes the world go round; causes the pulse to beat higher, the thrill of battle to rouse the sluggish blood, the brain to do ten times the work it can do when plodding along in emotionless tranquility.

"And in that possibility of enthusiasm," concluded the *New York American,* "lies the certainty of the future."

CHAPTER ELEVEN

The Path of Deliverance and Safety

OCTOBER 5–NOVEMBER 5

There she lies, the great Melting Pot—listen! Can't you hear the
roaring and the bubbling?
—From *The Melting Pot* by Israel Zangwill
Premiered October 5, 1908

It came with the darkening of the heavens, and the multitude
saw from the tower of the Times building shoot a ribbon of sil-
ver light.
—*The New York Times*, November 4, 1908

A ppearances to the contrary, the world did not stop in those last breathless days of the National League pennant race. It may have swerved a little, like one of Three-Fingered Brown's inscrutable curve-balls, but it did not stop. In America, the men who were running for president continued to ricochet through the states on special-service trains, stumping for votes. In South Carolina, a mob of whites attempted to break into a county jail and lynch a young black man they believed guilty of molesting a young white woman. In Kentucky and Tennessee, Night Riders continued to mount audacious attacks on their foes. Harry Thaw persisted in claiming his sanity. Orville Wright prepared to get up from his bed in Washington and go home to Dayton, while his brother, Wilbur, set new records for height and distance in Le Mans, France.

In another corner of the world, war nearly broke out in the Balkan states of the crumbling Ottoman Empire. Bulgaria claimed its independence from Turkey; Austria-Hungary declared its annexation of Bosnia-

Herzegovina; Turkey mobilized against Bulgaria; Serbians demanded war against Austria-Hungary; Germany declared its support for Austria-Hungary; British war ships steamed for the Aegean; and French diplomats pressed for peace. All of this occurred within the first two weeks of October, threatening to usurp the newspapers' attention to baseball. It was a dangerous dance of alliances and age-old enmities, in many ways a ham-fisted rehearsal for the opening scenes of World War I, which would begin in the Balkans six years later.

For the moment, anyway, the U.S. had no direct interest in the Balkans and stayed clear of its internecine quarrels. The attentions of the nation's diplomatic corps and military leaders were fixed, instead, on the impending drama beyond the opposite side of the continent, across the Pacific Ocean and near the eastern edge of Asia. There, on the morning of October 9, the Great White Fleet lifted anchor and steamed out of the Bay of Manila, turned north into the South China Sea, and headed to Japan.

Japan had been a long time coming on the fleet's itinerary. After leaving San Francisco in July, the battleships had bounced around the South Pacific, as if putting off an unpleasant chore. First Hawaii, then New Zealand, Australia, and the Philippines. The Manila stop had been something of an embarrassment for Americans and Filipinos alike—a cholera epidemic forced Rear Admiral Sperry to cancel shore leave for his men, much to the dismay of Manila's merchants—but otherwise the Pacific tour had gone off without a serious hitch. The Americans had been greeted as conquering heroes at every port.

Now, though, commenced the most delicate phase of the entire forty-three-thousand-mile cruise. In just over a week, the fleet would enter the Bay of Tokyo to begin a weeklong visit in Japan. News reports from Japan told of elaborate preparations for the Americans' arrival at Yokohama Harbor, but the fleet's commanders could not dismiss the possibility of an elaborate trap. Even barring sinister intentions on the part of their hosts, the pitfalls in Japan were numerous. A few drunken American sailors on the loose in Yokohama, a misunderstanding here, a racial slur there, and fragile relations between the two countries might descend into an abyss that would make the Balkans look like a teacup. This was the gamble President Roosevelt assumed when he sent the fleet into the Pacific ten months earlier. Now came the moment of reckoning.

*   *   *

President Roosevelt did not seem to be losing much sleep fretting over the fleet's visit to Japan. He threw himself into work and sport with his usual vigor in the early weeks of October. He played frequent games of tennis on the White House court, coached Republican campaign managers on last-minute tactics, met ambassadors from various nations, sat for his portrait and—simultaneously—composed a lecture on South American paleontology (for later delivery at Oxford University).

On the evening of Monday, October 5, accompanied by his wife and several cabinet members, he attended the Washington premiere of a new play by Israel Zangwill entitled *The Melting Pot*. The play told the story of a Russian-Jewish immigrant to America who learns to assimilate into the culture of his adopted country. The plot of Zangwill's play would soon be forgotten, but its title—*The Melting Pot*—became an instant catchphrase and lasting metaphor to describe America's capacity to absorb foreigners and synthesize them into patriots. The president led the applause when the play was done, then invited the playwright to lunch so that he could suggest a few improvements.

What time remained to Roosevelt after he was done with his duties as commander-in-chief, campaign strategist, paleontologist, and dramaturge, he devoted to planning his trip to Africa. This was a subject that came increasingly to preoccupy both him and the White House press corps. He wrote numerous letters that fall seeking counsel from Africa experts and ordering wares he would need. He examined guns and other hunting supplies that arrived at the White House by the wagonload, express from purveyors all over the world. He also applied for a hunting license. According to the *New York Times*, the license would permit him to kill as many lions, leopards, and crocodiles as he liked, in addition to a limited menu of other animals and birds:

Two male elephants, two rhinoceroses, ten hippopotami, twenty-one antelope, including two kudos, two gembok, and one bongo; two earth hogs, two earth wolves, ten chevrotains, mush deer, two colobi or other fur monkey, two marabou storks, two ostriches, two egrets, and one chimpanzee.

"I fairly dream of the trip," he'd written in late September to his friend the legendary English big-game hunter Frederick Selous. "I long to see the wild herds, and to be in the wilderness."

As details of the journey began to leak out that fall, the press called attention to its hazards. In the *New York World*, a writer well versed in the lethalities of Africa suggested the president would be effectively committing suicide by going on his hunt: if the fierce animals did not kill him first, the climate and diseases certainly would. "The African sun is death-dealing to highly excitable and full-blooded men," the writer warned, adding that men of the president's high-strung temperament were "subject to a special form of the bilious fever, which decomposes blood and causes death in twenty-four hours."

Before Roosevelt could hemorrhage to death in the jungles of Africa, though, he had to survive the jungles of Washington.

A Walk in the Park

On the afternoon of Saturday, October 10, Roosevelt went for a walk in Rock Creek Park. This was the capital's own little glade of wilderness that stretched along a valley through the northwest section of the city. Roosevelt had long enjoyed strenuous exercise in the park. Indeed, his Rock Creek rambles were legendary. Often dragging an entourage of cabinet members, diplomats, and various other Washington heavyweights in tow, he bushwhacked through thickets, scaled boulders and cliffs, and jumped into ice-choked creeks. Earlier in the year, he'd ended a hike with the French ambassador, Jules Jusserand, by undressing and jumping into the Potomac River for a swim. The ambassador stripped down and followed.

Not all were as game as Monsieur Jusserand. For true Washington heavyweights—the fat, the unfit, the deskbound—these hikes were grueling exercises in humiliation. A man unable to match the presidential stride could pretty well count himself down a few notches in Roosevelt's esteem, not to mention in the hierarchy of Nature. Ten years after the Spanish-American War, the president still seemed to judge a man by his readiness to charge up San Juan Hill should the need arise. (Roosevelt made a notable and curious exception in the case of William Taft, whom he judged so obese that all forms of exercise were ill-advised.)

On this particular afternoon walk, Roosevelt was accompanied by just one fellow hiker, Captain Archibald Butt. Archie Butt was a forty-three-year-old career army officer who served as the president's chief military attaché. The job combined the role of personal aide to the president, social secretary, informal adviser, and all-around glorified footman. Archie Butt's relationship with the president was newly minted but close, particularly for one so constrained by protocol. Since arriving at the White House the previous spring, Butt had come to develop a sort of schoolboy crush on the president, admiring all those qualities others admired in him (and some ridiculed), including his prodigious energy, his fervid intelligence, and his moral righteousness. Less obvious is what the president saw in Butt. Though a soldier, the captain hardly seemed the sort of man's man toward whom Roosevelt naturally gravitated. On the contrary, he was a putty-faced mama's boy, a lifelong bachelor who gave more attention to the cut and trim of his uniform and his collection of antique baubles than to weighty affairs of state. His value to Roosevelt was in his smooth southern decorum (he was a product of Georgia), his unlimited tact, and—probably most important of all—his utter devotion.

Whatever the precise reasons, Roosevelt enjoyed Butt's company and had recently started taking him into his confidence, speaking freely on matters of intellectual and personal interest—of growing old, of literature he enjoyed, of friends and light gossip. Butt worried about his own intensifying closeness to the president. "Sometimes his friendship almost frightens me," Butt wrote in one of his frequent letters to his beloved mother that fall; "when I am with him I become stampeded for fear that I may do something that if he knew he would not approve."

Roosevelt and Butt arrived deep in the park on the president's carriage shortly after four that Saturday afternoon. It would soon be dark, but the president dismissed the carriage, along with the two Secret Service men who had accompanied him. Only Butt would be there to protect the president from assassins. Though not a bodyguard in any official capacity, Butt had recently started carrying a pistol for exactly that eventuality.

The president led the way briskly into trackless brambles, "like a schoolboy, kind of dancing all the way," Butt wrote. They spoke as they walked. Butt posed questions and the president digressed across a wide spectrum of interests and opinions, being short of neither. As they discussed the baseball season, the president exhibited a keen knowledge of the just-concluded pennant race between Chicago and New York. The conversation turned to Edgar Allen Poe ("our one supereminent genius"), then to the career of George Washington ("the greatest man in our history"). Roosevelt ruminated about Washington's decision to refuse a third term, which he took as a model for his own conduct. "I could easily have persuaded myself that I was really needed to carry out my own policies," he told Butt. "Nine tenths of my reasoning bade me accept another term, and only one tenth, but that one tenth was the still small voice, kept me firm."

He did not mention the fleet's imminent arrival in Japan during the walk. Earlier in the summer, though, he had confessed his belief to Butt that war with Japan was, sooner or later, inevitable. "No one dreads war as I do, Archie," he'd told Butt, a surprising admission from a man often portrayed as a warmonger. "The little I have seen of it, and I have seen only a little, leaves a horrible picture in my mind." This was a side of Roosevelt that few people ever got to see, the man of probity and sensitivity. In a letter to his mother written a few days before the walk in Rock Creek Park, Butt had written: "I never did take much stock in the reports that he was constantly doing things without giving them thought—in other words, going off half cocked. . . . He is not as impetuous as he likes to appear."

Butt would have ample opportunity to reconsider this assessment during the course of their walk. Coming upon sheer rocks that jutted from the banks of the creek, the president scrambled up their steep faces and crawled over their narrow ledges, as Butt nervously followed. "Sometimes we had to pass ourselves along the outer faces of rocks with hardly enough room in the crevices for fingers or feet," wrote Butt. "My chief anxiety was for him. I felt that he had no right to jeopardize his health and life as he was doing."

Bad enough that Butt should suffer grievous injury trying to impress his boss. Worse that the president should get maimed or killed under his watch. Butt pursued Roosevelt through Rock Creek Park like a parent chasing a headstrong toddler—a toddler who happened to be president of the United States. Though Butt was much too discreet to suggest as much, the adventure was a kind of farcical nightmare. Around every corner the president found some new means of exterminating them both.

They came to a cliff that rose above the swollen waters of Rock Creek. The jagged ascent above the water was damp and slippery from the same recent rains that swelled the creek with swirling water. Shards of rock protruded from the face of the cliff. A fall might be fatal. Without hesitation, the president began to climb, as Butt watched with growing alarm from below. Just as Roosevelt was about to reach the highest point, Butt's greatest fear was realized: the president lost his grip and fell. "Had he swerved, his head would have been certain to strike some projection," Butt wrote later. "I stood paralyzed with fear. I could see what it would mean to have him meet with any accident of this kind."

Fortunately, the president managed to shove himself out from the cliff as he fell. The move saved his head from collision with the rocks. He plunged feet-first into the water below and sank to his shoulders. A moment later, he emerged on the shore, dripping and laughing. Whereupon, to Butt's horror, he immediately started back up the cliff. Butt knew the president well enough by now to realize there was no point in objecting.

This time, Roosevelt made it to the top without incident. Butt was so relieved by the president's survival that he scarcely paid attention to his own fears as he began to scale the rock. "I went over the ridge like a cat," he wrote proudly to his mother.

Walking on, they came to another deep pool of the creek. A White House aide had nearly drowned attempting to swim across this very spot earlier in the year. Advising Butt to swim "hard and straight," Roosevelt started across. Butt followed. They wrestled through the current and arose on the other side. On they walked through the dusky autumn woods.

Near the grounds of the National Zoo, within scent of the animals in their cages (no doubt their proximity made the president think of Africa), they came to one last steep cliff, this one rising about forty feet. Announcing that any attempt to climb it under such circumstances of darkness and dampness would be risky and probably impossible, Roosevelt began to climb. Butt found himself once again in the discomfiting position of watching the president of the United States risk his life. Not that there was much to watch from where Butt stood. The sun had set and the woods were dark. Roosevelt's body dissolved as he rose into the charcoal background of tree branches, rock, and sky, only his chafing feet and hard breathing indicating the route of his ascent. Then the president's voice called down. He'd reached the top.

"I could hear him calling from above not to attempt to follow," Butt wrote, "that it was too slippery and that it would be fatal to fall." Butt tried anyway. In the dark, he could not find footholds or handholds. Halfway up, he stopped, unable to climb further. He gingerly descended, then went around the cliff on a detour, where he met the president "coming out of the precipitous jungle like a bear, but laughing and evidently buoyed up by his prowess."

Together, the men climbed an embankment onto the nearest road. As no carriage awaited them, they walked in the cold dark toward Pennsylvania Avenue, soaking wet and splattered with mud. No doubt they made an interesting impression on those they passed, even on those who failed to recognize the stout bedraggled man in the Rough Rider hat as the president of the United States.

Japan

A few days into its voyage from the Philippines to Japan, off the north coast of the Island of Luzon, the Great White Fleet ran into the worst weather of the voyage. Monsoons brought ripping winds and high frothy seas, rougher than even the most experienced sailors of the fleet had encountered. Waves pounded the ships' hulls and crashed over their decks, tearing lifeboats from their moorings. Two seamen were swept overboard in the storm; one from the *Minnesota* was rescued with a well-thrown life buoy, another from the *Rhode Island* drowned in the swells. Rear Admiral Sperry ordered the ships to reduce speed from ten knots to eight knots and increase intervals from four hundred feet to eight hundred feet. For several miserable days the ships rolled and tossed as the men huddled under battened hatches.

When the weather cleared on October 15, the Americans, now a day behind schedule, got their first sight of the southern Japanese island of Kyushu—and of a squadron of three gray Japanese battleships approaching from the distance. The Japanese came not to harm them but to welcome them. After performing a series of maneuvers for their benefit, a sort of naval greeting dance, the Japanese warships turned and escorted the Americans along the eastern edge of Japan to Yokohama Harbor.

The fleet's arrival at each of its ports of call had been met with lavish and marvelous displays of affection. Nothing, though, came close to what awaited the Americans in Tokyo Bay on the morning of October 18. It was just dawn as the ships appeared, filing through the fog at the entrance of the bay, but already countless vessels were on the water to meet them, including seven large ocean liners, carrying a thousand passengers each. Anchored in Yokohama Harbor were sixteen Japanese warships, a companion for each of the American ships. These ships blasted their guns in welcome as the fleet approached. Fireworks dazzled the gray morning sky and bands played on the ships' decks. The mist lifted to reveal the hills surrounding the harbor. They were covered with people, tens of thousands of Japanese waving from hilltops and rooftops, more people, even, than had populated the hills of San Francisco back in July. "The spontaneous enthusiasm of the populace," reported the *New York Herald*, "is unbounded and unprecedented."

The following morning, Sperry and a number of high-ranking American officers boarded a special train to Tokyo, where they would spend the week in a round of garden parties and receptions, dinners and balls. The train was draped in flags and filled with flowers. The officers had remained skeptical of the Japanese's intentions for the first twenty-four hours of the visit, but along the eighteen miles of track to Tokyo their misgivings melted away. Passing through villages of thatched-roofed homes, they were hailed by an almost continuous ovation of schoolchildren who lined the track, waving American flags and cheering. "The journey was, in fact, one long procession through a lane of waving flags," described the *New York Times*. Upon their arrival in Tokyo, the American officers walked through a corridor of ten thousand more schoolchildren who serenaded them with "The Star-Spangled Banner."

On Tuesday, the officers visited the Imperial Palace for an audience with the emperor, then sat for a luncheon. Everywhere they went, a chorus of voices reminded them how dear America was to the Japanese. "It is painful to every Japanese to be aware of the strange notion current in some parts of the United States that we are nourishing sinister designs against the land of Lincoln," the mayor of Tokyo told them. "The Japanese nation asks you to convey the message that the Japanese believe a war between Japan and America would be a crime against the past, present and future of the two countries."

As the officers passed their days and nights among the aristocracy of Tokyo, large "liberty parties" of American bluejackets took to the streets of Yokohama and Tokyo accompanied by Japanese sailors and translators. They found friendly crowds on bustling streets decorated with American flags. Local newspapers ran English editions for their perusal and streetcars were free to them by official decree. "The American uniform," reported the *New York Times*, "is the open sesame."

"Japan has captured the American fleet," teased the *Herald*. "That captivity is wonderfully agreeable to every officer and man."

Rear Admiral Sperry and his fellow officers returned to Yokohama on Friday, after five days of festivities. By the time the fleet lifted anchor and steamed out of the harbor early on the morning of Sunday, October 25, it was clear the visit had been a total success. "There is no parallel in history that we recall to the reception of the fleet of the United States battleships in Japanese waters and the warmth and unquestionable sincerity of the greeting of our sailors on shore," the *New York Times* gushed in an editorial. As for war, the very idea now seemed preposterous. "No nation teaches its children to sing the songs of a people for whom it has unfriendly feelings."

A Ribbon of Silver Light

Theodore Roosevelt turned fifty on October 27, two days after the fleet left Japan. All through the day couriers arrived at the White House bearing messages of congratulations. Horses clopped up the drive pulling express wagons overflowing with flowers and gifts. Diplomats filed in to deliver their best wishes. Finding the president indisposed, most left their calling cards and departed without seeing him.

Roosevelt seemed intent on ignoring his half-centennial. After a quick cabinet meeting in the morning, he retired to his office to work. Just after 4 P.M., he emerged from the main entrance of the White House dressed in a khaki riding suit, Rough Rider hat, and boots and spurs. He was delivered by carriage to Rock Creek Park, where he treated himself to a late-afternoon ride on horseback through the park and into the countryside beyond.

Really, the president could not have asked for a better birthday gift than the one given him by the fleet. Its success in Japan had to be the sweetest kind of vindication. What many of Roosevelt's opponents had condemned as heavy-handed and tactless had turned out to be deft and diplomatic. A potentially explosive situation had been defused, a threat had been neutralized, and America's role in the Pacific had been bolstered. "It is difficult to overestimate the future importance of this event," stated the *New York Herald*, "when it is remembered that the United States is the predominant Pacific naval power as long as the Atlantic fleet remains in this ocean or continues to be able to repeat its quick transfer from ocean to ocean." In keeping with his theory of the

Big Stick, Roosevelt had added luster to America's reputation as a global power without ordering a single shot fired during his presidency.

Roosevelt had plenty of reasons to feel proud that fall. The domestic economy had improved remarkably since the previous year. Indeed, newspapers were reporting that signs of the 1907 panic were gone without a trace. As for the success of Roosevelt's reform polices—those same policies his foes had blamed for causing the 1907 panic—the platforms of the two men running to replace him gave some indication of their popularity. Both William Howard Taft and William Jennings Bryan were promising to continue more or less in Theodore Roosevelt's footsteps.

Of course, it was no small matter to Roosevelt which man succeeded him. He had devoted a good deal of energy to ensuring Taft's victory, more energy, possibly, than Taft himself. Bryan had not made the job easy. The Democratic candidate had cobbled together a formidable alliance of conservative southerners, midwesterners, liberal easterners, and labor unions to form his base. And while Taft moved through the campaign at a lumbering pace, at least at the start, Bryan, the Great Communicator, burned across the country delivering his famously passionate oratory at every whistle stop. In one four-day stretch in Nebraska in early October, for example, he'd averaged an astonishing twenty-one speeches a day. Taft may have had Roosevelt in his corner, but Bryan had the support of a good percentage of the southern and midwestern populace. No one could count him out.

But then, in mid-September, the Republicans got a lift from an unexpected quarter. The newspaper publisher and political aspirant William Randolph Hearst began printing letters that had been purloined several years earlier from an official at Standard Oil, the most despised trust in America. Hearst had been holding on to this toxic correspondence since acquiring it, awaiting a timely opportunity to spring it on the public. Hearst's motive for producing the letters in the autumn of 1908 was evidently to burnish the standing of his newly founded political machine, the Independence Party. The more immediate effect, though, was to severely damage William Jennings Bryan's chances. The letters implicated a man named Charles Haskell, who happened to be the governor of Oklahoma—and who happened to be, more to the point, William Jennings Bryan's campaign treasurer. The letters suggested that Haskell had done favors for Standard Oil in exchange for financial contributions. While nothing in the letters implicated Bryan directly, the damage was done by association.

Never one to miss a chance, Roosevelt decided to "put some ginger" into Taft's sleepy campaign. Throughout October, he excoriated Bryan and the Democrats for their moral shortcomings. As a matter of fairness, the charge was specious; as politics, it was highly effective. By the eve of the election, the Democrats were in obvious trouble. Straw polls conducted by newspapers pointed to a Taft victory. More tellingly, perhaps, the betting parlors of Chicago were giving five-to-one odds for Taft. All this had to be gratifying to Roosevelt.

Not so gratifying was the snide tone of his constant critics. The editorial page of the *New York Times,* for instance, dismissed Roosevelt's involvement in Taft's campaign as intrusive and possibly counterproductive. "Many sober friends of Mr. Taft, and even of Mr. Roosevelt, wish the President had kept more quiet," contended the *Times.* "There is an impression, especially in the East, that Mr. Taft would have been stronger if public attention had been allowed to concentrate upon him, without the distractions due to the demonstrations from the White House . . ."

In the end, "Whether he has done Mr. Taft more good or harm is a question on which there is room for difference of opinion everywhere except in the mind of Mr. Roosevelt."

William Taft ended his campaign with a swing through his home state of Ohio. After waking in western New York on November 2, he traveled by train through Cleveland to Youngstown, with numerous whistle stops between. His last speech, made on the eve of the election in Youngstown, was the 418th he'd delivered in the last 41 days—not an oratorical achievement of Bryanesque proportions, but an admirable exertion nonetheless, particularly in light of his lethargic start.

As for Bryan, he wrapped up his campaign in Marysville, Kansas, before heading home to Lincoln, Nebraska. In contrast to Taft's confident tone, he sounded almost wistful as he closed his third and final run for the presidency. "I am now forty-eight years old," he told the crowd in Marysville. "I know not what the future has for me. I know not whether it is the people's wish that I shall be their spokesman in the White House, or continue to perform the work which I have tried to perform as a private citizen, but I have not lived in vain."

The president was in good spirits that day. "We've got them beat to a frazzle," Roosevelt announced on the eve of the election, beginning to savor the victory he now believed certain. When the President used the first-person plural—"we've"—there was no doubt he considered the victory substantially his own.

Just before midnight of November 2, Roosevelt and his wife boarded a northbound train at Washington's Union Station to begin the long journey to Oyster Bay. Like many thousands of Americans, the president would travel hundreds of miles to cast his vote. A direct nighttime run up the Atlantic seaboard took the Roosevelts to Jersey City, where a tugboat met them and ferried them across New York Harbor to Long Island City. There they were joined in the early morning by their twenty-one-year-old son, Theodore Jr., who had come down from Connecticut to cast his first vote in a presidential election. The family traveled together through Long Island to Oyster Bay. As Mrs. Roosevelt waited in the carriage, Theodore Sr. and Theodore Jr., accompanied by a mob of press and well-wishers, entered Fleet's Hall to cast their ballots. Following a quick stop at Sagamore Hill, they were all back on the train by eleven o'clock in the morning, to return just as they had come.

Election days were different then. They were grander occasions, at once more solemn and more festive. Only men voted—women's suffrage was still a dozen years away—but of those who could vote, more than 65 percent went to the polls. This was at a time when voting might require a day's hard journey. It was a time, too, before television and radio, so news of results had to be sought out beyond the home. Election nights were therefore social occasions, group occasions, especially in towns and cities, where people poured into the streets to await news and to celebrate or mourn. If election *days* were for practicing a sacred rite of American citizenship, election *nights* were for enjoying the all-American rights of free assembly, riotous enthusiasms, confetti-tossing, and whatever else that followed.

The crowds in New York were larger than ever on the night of November 3, 1908, a fact possibly explained by the balmy, clear weather, or perhaps, more simply, by the fact that the population of New York was several hundred thousand greater than it had been four years earlier. Newspaper reporters strained for hyperbole. The *New York World* probably took the prize when it described the crowd at City Hall Park as larger "than has at any time in the earth's history taxed that particular portion of the earth's crust."

From City Hall to Harlem, Broadway teemed with people. The most intense concentrations naturally occurred where the city's chief newspapers were headquartered—at Newspaper Row downtown, at Herald Square and Times Square uptown—and where the news, posted and illuminated, was likely to be freshest. No place packed a larger crowd that evening than Times Square. The *New York Times* estimated it at fifty thousand, about the same as had attended the start of the New York–to–Paris race in February. Several hundred mounted and regular policemen under the command of Inspector Max Schmittberger—the same Max Schmittberger who had led the troops at Union Square in March—sought to control the vast sea of humanity by keeping it in perpetual circulation, a moving body presumably being more pliable than a body at rest. The *Times* reported that Schmittberger's men pushed the crowd on the east side of Broadway *northward* and pushed the crowd on the west side of Broadway *southward*—or was it, rather, as the *World* reported it, *southward* on the east side and *northward* on the west side? The details mattered little in the end. Despite Schmittberger's best efforts, the northward crowd and southward crowd soon spilled across Broadway and merged in the great swirling melting pot that was Times Square on election night.

Many in the crowd came out for the serious purpose of gleaning the results; others treated the night simply as an excuse to celebrate, no matter who won. The goal of this latter group, explained *Harper's Weekly*, "was to avail itself to the limit of its freedom and to exasperate every fellow human creature it encountered who did not happen to be in its own regardless mood." Like New Year's Eve and Independence Day, election night was one of those cyclical moments when the city abandoned itself to its most frolicsome, childish, and democratic instincts.

The *Times* was the first newspaper to project a winner. The projection came quite literally at 6:10 P.M. in the form of a powerful beam cast by a searchlight from atop the Times skyscraper. "It came with the darkening of the heavens," reported the *Times*, "and the multitude saw from the tower of the Times building shoot a ribbon of silver light. It shot through the gathering gloom to the north and pierced the skies." By previously published code, a beam pointing to the south would have meant Bryan was the victor. This northward beam meant Taft.

The *New York Herald* followed with its own searchlight projection at five minutes to seven. The *Herald* light, set atop the nearly completed

Metropolitan Life tower, flashed on forty-five minutes later than the *Times's*, but its source was hundreds of feet higher and its beam reached dozens of miles farther—sixty miles from the center of Manhattan on a clear night, and into three different states. "Never before has an attempt been made to indicate election night returns from so lofty an eminence," bragged the *Herald*. The searchlight was so lofty, in fact, that despite calm atmospheric conditions at street-level, its operators, perched almost seven hundred feet overhead, stood in the midst of a heavy gale. Holding tight to the building's steel frame, they turned the searchlight to the north.

Two hours later, when the *Herald* swung its searchlight westward to indicate Charles Evans Hughes's reelection as governor of New York, the wind atop the Metropolitan Life tower was so severe the operators had tied themselves to the steel like sailors lashed to the mast in high seas. From far below, over the rush of the wind, they could hear the siren call of the crowd, the horns and bells, the clamor that lasted long after the returns were all in and the results were all known.

Taft had won, and he had won big. He received only 51.6 percent of the popular vote next to Bryan's 43 percent (Eugene Debs and various other

The searchlights of the New York Times building point to the north and west to indicate victories for Taft and Hughes.

alternative candidates took the rest), but in the electoral college he won by a landslide, 321 votes to Bryan's 162. Mirroring the code of the newspaper searchlights, Taft's voters were oriented to the north, Bryan's to the south. A map of Taft states and Bryan states looked remarkably like a map of Union states and Confederate states half a century earlier.

At midnight, President-elect Taft went out to the front porch of his brother's house in Cincinnati to address a crowd of supporters waiting on the lawn. "I say that so far as I can pledge to you all the energy and ability that in me lies shall be used to make the next administration a worthy successor of that of Theodore Roosevelt, and beyond that I claim nothing higher." He neglected to add that he first intended to devote his energy and ability to playing golf for three weeks in Hot Springs.

Bryan, at his home in Lincoln, Nebraska, was gracious in defeat. Or maybe he was just in denial. As returns had filtered in throughout the day, he'd kept an upbeat attitude. He performed a pianola recital in the drawing room for the diversion of the newspaper correspondents and other visitors at the house, then graciously invited everyone for dinner. He behaved all along as if he believed he was going to win; and when it became clear that he would not, he politely excused himself and retired to his room.

Probably the happiest man in the country was sitting in the White House that night. President Roosevelt had returned to Washington from Oyster Bay just in time to host a small victory party. The news coming in over the White House cables was even better than he'd expected. Still, he did not stay long to celebrate. As his wife and guests socialized below, he went upstairs to sit alone and read. What he read about is not recorded. South American paleontology? Africa? One suspects that whatever the topic, and even allowing for his immense powers of concentration, he occasionally lowered the book to gaze across the room and reflect for a few moments on all he had accomplished—and all that he would soon be leaving behind. Perhaps a small filament of regret flickered in the back of his thoughts. A suspicion that life might never be so good again.

The next morning Roosevelt was still in high spirits, but at noon he turned serious long enough to interpret the election results for the White House press corps. "The nomination of Mr. Taft was a triumph over reactionary conservatism," he told the reporters, getting in a dig at the old guard of his party, "and his election was a triumph over unwise and improper radicalism." Even Roosevelt seemed to accept that Taft's steady centrism would be a welcome break from the Sturm und Drang of his own administration. If so, he anticipated the sentiments of his constant critic, the *New York Times*. "The people have voted for peace," declared the *Times* in an editorial. "They have voted for peace through the ending of agitation, tumult, and vociferation." In Taft, they had found "the path of deliverance and safety."

The *Times*'s implication that the election was a repudiation not only of Bryan, but of Roosevelt, strained credibility. Taft's entire appeal to voters had been that he *was* Roosevelt. Americans elected Taft not in spite of Roosevelt, but *because* of Roosevelt. Still, the *Times* was right about one thing. The White House would be a quieter place when he was gone.

Two days after the election, November 5, on a cold and windy afternoon, Archie Butt went horseback riding with Roosevelt in Rock Creek Park. He had not seen much of the president since their tramp through the woods in early October. At the end of the month, Archie's mother—his beloved mother to whom he had written so many letters about Roosevelt—had passed away and he'd returned home to Augusta, Georgia, to bury her. He was devastated by her death. Typically selfless, he offered to quit his work at the White House lest his grief intrude on his duties. The president and Mrs. Roosevelt begged him to stay. Perhaps the president hoped this ride in the woods would take Butt's mind off his mother's passing.

As it happened, though, their conversation turned to death almost as soon as Butt and the president trotted into the November woods. The men had been lightly discussing Taft, specifically whether a man of such size should ride horseback. "I found that the President agreed with me that he should not attempt to ride, that it was dangerous for him and cruelty to horses," wrote Butt later in a letter to his sister-in-law. Roosevelt remarked that Taft would be better off avoiding all forms of exercise, concentrating his energies, instead, on the business of being president. "If I were Taft," said Roosevelt, "I would content myself with the record I was able to make in the next four years or the next eight and then be content to die." More important to live well, Roosevelt suggested, than to live long. "I am ready to go at any time. Certainly the fear of dying would not deter me from doing what I wanted to do.

"I do not know what the future has in store for me, but I am ready to rest my case here," said the president. "Or after I have had a little fling in Africa."

CHAPTER TWELVE

The Modern Definition of Life

NOVEMBER 5–DECEMBER 31

> *The modern definition of life is a power to gather the material*
> *of the universe to ourselves, to make it our own, and use it*
> *under the control of a well-trained will.*
> —*The Independent*, December 31, 1908

About the same time Theodore Roosevelt and Archie Butt were trotting on horseback through the woods of Rock Creek Park that Thursday afternoon in Washington, Wilbur Wright was setting down his dessert fork at a banquet table in Paris. It was late evening, the hour when dinners end and digestifs are poured, cigars are lit and speeches commence. Because the speeches were in French, Wilbur Wright could not catch every word but he got the gist. When his name was spoken—the French version, *Velbur Reet*—it was his cue to rise. A burst of applause rang from the three hundred men seated in the great hall—la Salle des Fêtes—on the Place de la Concorde. Six hundred eyes gazed on him.

The Wilbur Wright who stepped up to the dais that November night was a different man than the one who had walked off the ship at Le Havre back in late May. An autumn of lunches and banquets had added almost ten pounds to his frame, lifting him for the first time in his life above 150 pounds. Months of outdoor work, of manhandling the airplane in sun and wind, had turned his face ruddy, scoring lines at the corners of his eyes and adding sinew to his forearms and shoulders.

And then there were the less definable changes. He had always projected the quiet charisma of a confident man who keeps his own counsel, but he was a more substantial and imposing figure now. He was, indeed, among the most famous men in the world. Visiting dignitaries, rich businessmen, and military officers came to pay him court at Camp d'Auvours. Some wished to talk business, others to satisfy their curiosity, others to indulge in the ultimate experience—and claim the unbeatable boast—of having flown with Wilbur Wright.

Those whom Wilbur chose for this honor climbed in among the cables and levers, tucking their bodies into the little seat at the front of the lower wings. Wilbur slipped nimbly in beside them. One of his assistants drew a trip wire across their chests; should the plane crash, their bodies would thrust forward, pulling the wire and cutting the engine. They must have all felt, in the moment before takeoff, a surge of regret: the realization of what exactly they had signed onto, the recollection of Lieutenant Selfridge dashed to his death (they had no doubt seen the photographs), the mounting instinct to bail. But already it was too late. Everybody was waving, smiling. The motor screamed to life. The propellers clattered behind. And then they were shooting down the rail, gathering speed, rising above Camp d'Auvours, over the crowds below, the roofs of the buildings and sheds of the artillery range, the tops of trees, as countryside unfurled beneath them.

These two-man flights were no longer lurching little hops, as they had been as recently as September. They were sustained aerial excursions, many approaching an hour in length, some longer. Nearly every week now Wilbur flew longer or higher or farther than anyone had ever gone before. On October 10, he set an endurance record for a two-man flight, remaining aloft with a passenger for one hour and ten minutes. A week later, he set a new record for altitude, flying to 230 feet.

Several of Wilbur's flights that fall earned him prizes, which were welcome, and money, more welcome still. The Wrights had yet to amass anything like a fortune, but the more Wilbur flew, the richer their prospects. The French partnership was deluged with offers to buy stock. Negotiations with the Russians, the English, the Italians, and the Germans were under way, while the American navy appeared ready to make an offer, pending the resumption of Orville's flight trials at Fort Myer.

All of the flying and deal-making took a toll on Wilbur. In those creases around his eyes, the men in the Salle des Fêtes might have detected exhaustion. Wilbur's work had been unremitting. Between flights, he had to contend with business while handling an impossibly large correspondence (three-quarters of which he'd never answer), not to mention overseeing the mechanical maintenance of the plane that flew between him and death. Especially now that he was frequently carrying passengers, and in light of Orville's crash, he had to be very careful. For a Wright to kill another passenger would not only be tragic; it would be fatal to the brothers' reputation.

With so many responsibilities and obligations pressing on him, it's no wonder Wilbur occasionally yearned to put all this aside and return to a simpler time. "How I long for Kitty Hawk," he wrote to his friend Octave Chanute. His fans at Camp d'Auvours may have looked to mechanical flight as a medium of transcendence, but Wilbur found his own escape on a bicycle. He rode away from the crowds and pedaled into the countryside. Children doffed their hats to him as he passed, calling out "Bonjour, Monsieur Wright." ("They are really almost the only ones except close friends who know how to pronounce my name.") Soon he was in the woods, unrecognized by anyone but the birds. Under the trees, his mind wandered back to Kitty Hawk, or to his family's home at 7 Hawthorn Street in Dayton. "I had hoped to be in America by Thanksgiving Day," he wrote to Chanute. "But now I only hope for Christmas."

But Wilbur would not go home. Not for Thanksgiving; not for Christmas or New Year's, either. There was simply too much to be done in Europe. For the next several months, his relationship with those closest to him would be strictly epistolary. Instead of the love of family, he would have the adulation of strangers. He always insisted that this adulation include Orville. As he wrote to his father, "[I]t would be impossible for me to accept an honor which Orville could not equally share." The prize he came to accept now, on this October night in la Salle des Fêtes, was for Wilbur and Orville *both*.

"For my brother and myself, I thank you for the honor you are doing us and for the cordial reception you have tendered us this evening," Wilbur began. He spoke just briefly. ("I know of only one bird—the parrot—that talks," he'd informed an audience earlier in the autumn, "and it doesn't fly very high.") He admitted that as recently as 1901, he had been doubtful not only of his and his brother's chances for success, but of the future of flight altogether. "I said to my brother, Orville, that men would not fly for fifty years. Two years later, we ourselves were making flights."

Since then, he told the audience, he'd made it a point to avoid making predictions about flight. "But it is not really necessary to look too far into the future; we see enough already to be certain that it will be magnificent."

War in the Air

The magnificent future was a bell rung frequently in America that autumn of 1908. The Wrights' success made it appear more dazzling than ever. "Is it possible that at this very moment we are living in the dawn of the veritable Golden Age?" wondered *Harper's Weekly* in mid-November. "The principal problems of aerial navigation seem to have been solved, and there is little doubt that before five years have elapsed swift and reliable airships will be as plentiful as swift and reliable automobiles are today." *Harper's* compared the Wrights' achievement to that of Christopher Columbus in 1492. Columbus had discovered a new geographic hemisphere; the Wrights had granted humans "access to every breathable portion of the earth's atmospheric envelope."

Harper's devoted a great deal of attention to aerial navigation. In another issue that fall, the magazine noted that some of the city's tall buildings "seem already constructed for the coming flying-machines, with projecting parapets to their flat roofs, on which one might imagine the aeroplanes hooking themselves in temporary anchorage." In addition to the convenience of skyscraper docking, *Harper's* speculated, aircraft promised health benefits, since the air was cleaner up high. Did not Princess Marie of Romania live in a tree fort atop a pine forest for exactly this reason? Perhaps in the near future, rather than send invalids to health resorts far away in the mountains, doctors might prescribe a few hours aloft "at an altitude adapted to the case at hand."

But even as *Harper's* pondered the wonders of the aerial age to come, its editors considered less salubrious possibilities. In mid-October, for example, the magazine ran a drawing by illustrator Vernon Howe Bailey entitled "The Next Step." The drawing shows Wright Flyers swooping through the air, evidently attacking the battleships below. Puffs of smoke rise from the ships. The sea appears to boil and steam around them.

That airplanes would find their chief utility in military rather than civilian life, in war rather than transit, was an assumption shared by most people who had an informed opinion on the subject. "It may as well be recognized that the aeroplane can never be seriously considered as a means of transportation on any extended scale," *Scientific American* stated. "The present indications are that a single machine can never hope to carry more than two or three passengers . . . Undoubtedly the greatest field of usefulness will be in military operations."

Depictions of what aerial warfare might look like were common in newspapers and novels. None were more vivid than those in H. G. Wells's *The War in the Air,* published in New York in October 1908. Wells's novel told the story of a young Englishman, a "vulgar little creature" named Bert Smallways, who ends up, by virtue of coincidence and his own idiocy, mistaken for a great aeroplane inventor. Taken captive by the German army, he finds himself sailing over the North Atlantic in a giant Zeppelin-like dirigible as part of a surprise attack on America. When the Germans come upon America's Atlantic naval fleet (or what remains of it, since most of the fleet is in the Pacific on its world tour), the dirigibles, along with a squadron of Wright-like biplanes called *drachenfliegers,* attack from above, killing every last man on board and sinking every vessel, including the American flagship, the *Theodore Roosevelt.*

The War in the Air was a farce, but it was a deadly serious and chillingly prophetic farce. Wells not only highlighted the currents that would pull Europe into World War I a few years later—megalomania, nationalism, an obsession with the toys of war—but grasped the way air fleets would change the nature of war forever. Though some in the military were still speaking of aircraft primarily as tools of reconnaissance and surveillance, Wells grasped how they would make their chief contribution: as delivery systems for armaments—as bomb droppers, city wreckers, human killers. This was something new. True, navies had trained their guns on urban ports in the past and armies had scorched their way through cities, but those campaigns had been undertaken in service to the main objective of war, which was to kill the opposing military. Air war would be different. Air fleets, singularly equipped to wage surprise attacks on urban populations, would treat civilian casualties not as accidents of war, but as a key strategic element of it.

In Wells's story, New York is the first city to experience the full horror of aircraft—"the first of the great cities of the Scientific Age to suffer by the enormous powers and grotesque limitations of aerial warfare," as Wells puts it. The German air fleet sails in over Manhattan and begins laying waste to the city. The Brooklyn Bridge is felled, Wall Street set ablaze. The Germans then sail up Broadway, releasing bombs, "dripping

Broadway under aerial attack in H. G. Wells's 1908 novel *The War in the Air*.

death," as Bert peeks through the portholes at the exploding city below. "And so Bert Smallways became a participant in one of the most cold-blooded slaughters in the world's history," writes Wells, presaging not only World War I, but the air campaigns of World War II, of Pearl Harbor, of London and Dresden, of Nagasaki and Hiroshima:

> He clung to the frame of the porthole as the airship tossed and swayed, and stared down through the light rain that now drove before the wind, into the twilight streets, watching people running out of the houses, watching buildings collapse and fires begin. As the airships sailed along they smashed up the city as a child will shatter its cities of brick and card. Below, they left ruins and blazing conflagrations and heaped and scattered dead; men, women, and children mixed together as though they had been no more than Moors, or Zulus, or Chinese. Lower New York was soon a furnace of crimson flames, from which there was no escape.

New York, city of steel of the golden age—melting pot of the world— melts into a heap of smoldering metal.

Last Licks

Most Americans were not overly concerned about getting bombed as they tucked into their Thanksgiving feasts that November. Aerial warfare remained safely confined to the pages of fiction. Peace reigned near and far. The Balkan crisis had subsided. The Great White Fleet was steaming triumphantly toward the Suez Canal, bound for home, having been received with adoration at every port of call. Nowhere had the reception been warmer than in Japan. Tensions between the U.S. and Japan had dissolved into mutual veneration. The terms of this happy friendship were soon to be made official. A few days after Thanksgiving, on November 30, Secretary of State Elihu Root would sign an accord with Japan's ambassador to Washington, Takahira Kogoro, that guaranteed respect for each other's territorial interests in the Far East and signaled general good will between the two countries.

Theodore Roosevelt's mood that November reflected the calm of America's global affairs. As President-elect Taft passed the month hitting golf balls in Hot Springs, Virginia, the president appeared to be mentally loosening his grip on the helm of the nation. He withdrew a little into himself and his work. Archie Butt noted that he lacked his usual vigor on the tennis court, as if his mind were preoccupied. For several weeks, the newspapers were devoid of colorful Roosevelt anecdotes.

The president spent Thanksgiving at the White House with his family and a few friends, including Archie Butt. After a dinner of turkey and suckling pig, Roosevelt and Butt retired to the den. The men later returned to the women to help them put together a jigsaw puzzle. PRESIDENT HAS QUIET DAY, ran the headline in the *Washington Post*. Given the man, this *was* news.

With the arrival of December, the truth of Roosevelt's quiescence became clear. He had not been slowing down; he had been husbanding his energies for a final ferocious burst of activity before the year ran out.

His return to form began on December 8. This was the day he delivered his last annual message—his valedictory State of the Union address—to both houses of Congress. The whopping document was distributed by couriers to the Capitol and read aloud by clerks.

In part because the president was now a lame duck, and therefore his words no longer mattered much to Congress, and in part because he had

written so confoundingly many of them—about twenty-one thousand—
the congressmen and senators did not initially pay much attention to
Roosevelt's message. Both chambers were nearly empty as the reading
clerks droned through it. Those few in attendance chatted among themselves, wrote letters, and scanned newspapers.

Page after page, the message promised little in the way of news. Most
of it was boilerplate Roosevelt: more rights for labor, more oversight of
trusts, less judicial tampering, better child labor laws, better protection of
natural resources—a stew of moderately progressive ideas. Some of these
ideas remained abhorrent to the old guard conservatives, but after two
terms of Roosevelt none qualified as surprises, and the calm tone in
which they were reiterated was almost lulling. So lulling, in fact, that most
of those who were present failed to notice the stinger when it came. It was
buried about fourteen thousand words into the message.

The president had been puttering on for some time about inland
waterways. From this, he had gradually shifted to a discussion regarding
the newly established Secret Service Agency. The subject might have
pricked up a few ears, had any still been paying attention.

Congress had recently passed an amendment limiting the Secret Service's jurisdiction. Congressmen who supported the amendment
defended it as a means to prevent the growth of an insidious secret spy
agency, "a black cabinet," within the government. To the president—as
he now made clear—the amendment was simply an attempt by congressmen to protect their own hides from peering eyes. "The chief argument of the provision," Roosevelt wrote in his message, "was that the
Congressmen did not themselves wish to be investigated by the Secret
Service men. . . ." He pointedly added, "I do not believe it is in the public interest to protect criminals in any branch of the public service."

By the usual standards of Roosevelt rhetoric, the accusation was far
from blistering. But the president must have known it would raise congressional hackles—would raise them, that is, just as soon as the dozy
legislators realized they had been insulted.

That night, the president, evidently in high spirits, invited a hundred or
so guests to the East Room of the White House to watch a movie with
him. The movie's footage had been shot in the far west of America and
showed a friend of the president, a United States marshall from Oklahoma named John Abernathy, hunting wolves with his bare hands. After

capturing the wolves, Abernathy would subdue them by shoving his fist down their throats. "That's one of the finest sports of the world," the president called out. Baiting congressmen was surely another.

By the next morning, members of Congress had been alerted to the offending section of Roosevelt's message and had worked themselves up into a lather of indignation. The president had essentially accused them of criminal conduct; it was an outrage, a calumny that demanded a strong rebuke. "Nothing that Mr. Roosevelt has done in his seven years in the White House has so stirred up members of Congress," reported the *New York Times*.

Congress's angry reaction only added more fuel to Roosevelt's fire. A day after his annual message, he launched an entirely new attack upon an entirely different adversary. This time it was newspaper publisher Joseph Pulitzer. The *New York World*, Pulitzer's flagship newspaper, had recently dredged up an old charge against the president, suggesting that financial improprieties had occurred in the deal to fund the Panama Canal early in his administration. Now it was the president's turn to be indignant. On December 9, Roosevelt asked the Justice Department to pursue libel charges against Pulitzer. "I do not know anything about libel law, but I should dearly like to have it invoked about Pulitzer," he wrote to U.S. Attorney Henry Stimson. "Pulitzer is one of those creatures of the gutter of such unspeakable degradation that to him even the eminence on a dunghill seems enviable."

Both of the fights Roosevelt picked that December would eventually fizzle out, but only after much ado. Congress would vote to rebuke Roosevelt early in 1909, but its actions would mean little by then since Roosevelt would be nearly gone from office; the United States would sue Pulitzer for libel, but would ultimately lose the case before the Supreme Court. In the meantime, anyone who worried Theodore Roosevelt might withdraw quietly into presidential retirement knew better now. "Cannot a plan be devised to keep Future Brother Roosevelt in this country?" implored *Harper's Weekly* on December 19. "We shall miss him terribly. Who will be left to supply us with exciting news after he goes away from here? Mr. Taft is a good and true man, and promises to make a good and true president. But he has no real knack for daily novelty."

Christmas

The Christmas of 1908 was a splendid contrast to the dismal holiday of 1907. Pocketbooks were flush and shops were crowded to bursting. Newspapers around the country reported that local merchandise was moving at record volume. American-made toys, newly electrified to suit American boys' whims, were attracting more buyers than European-made toys for the first time. Four million Christmas trees were being felled to decorate Americans' homes—so many trees that nature-lovers worried the forests were being depleted for the sake of holiday merriment. (Gifford Pinchot, chief of the Bureau of Forestry, assured his countrymen they needn't worry.) Still others fretted that an excess of gift-giving was distorting the true meaning of Christmas. "It is in the American temperament to make expensive gifts merely because it is the fashion to do so," wrote the author William Dean Howells in the *New York Times*. "It is to be hoped that we will return to a simpler and saner attitude toward Christmas."

But even Mr. Howells must have appreciated the booming economy that supported such generosity. Hardly more than a year had passed and the recovery from the 1907 panic was all but complete. Between December 25, 1907, and December 23, 1908, stock averages on Wall Street had risen by 30 percent, while wages had maintained their pre-crash highs. Unemployment, up to 8 percent during the crash, was almost back to its precrash lows of 3 or 4 percent.

"What has happened is such recuperation as is amazing," the *New York Times* gushed in an editorial. Other newspapers emphatically agreed. PROSPERITY ONCE MORE REIGNS IN THE BLUE GRASS, blared a headline in the *Lexington* (Kentucky) *Herald*. ERA OF SPLENDID PROSPERITY, echoed the *Charlotte* (North Carolina) *Observer*.

American farmers had harvested hundreds of millions of dollars more from crops in 1908 than in any previous year, and had exported more agricultural product than any other country in the world. American industry was growing even faster than American agriculture. Between the start of the century and the end of 1908, the nation's coal production had nearly doubled, its steel production had *more* than doubled, and its oil production had risen 150 percent—and the future promised only better. Fifty thousand automobiles, more than a quarter of all those on the road at the end of the year, had been manufactured in 1908. Very soon, hundreds of thousands—*millions*—of these would be on the roads, powered by gas-thirsty internal combustion engines. These same automobiles would require untold amounts of rubber, glass, and steel—all strong American industries that were on the verge of great expansion.

J. P. Morgan, whose U.S. Steel trust was the largest steel producer in the world—and would benefit enormously from the automobile's boom—articulated his own optimism for the country in a speech he delivered in Chicago on December 10. "Any man who is a bear on the future of this country," he told his audience, "will go broke." His words would be the mantra of bulls for years to come.

But the achievement of America was much greater than mere prosperity. A newspaper in Columbia, South Carolina, the *State*, looked back on the year of "Noteworthy History" and concluded that "America and American principles had an especially potent influence throughout the entire civilized world" in 1908. Both the "Globe-Encircling Fleet" and the "Conquest-of-the-Air" had "brought the United States into the foreground of international affairs more prominently than in many a year past." The *State* believed that these American triumphs had inspired oppressed people around the world, from Persia to Turkey to China, to stand up against tyrants and demand American-style liberty.

The *State* may have been giving America too much credit, but there was no denying the nation's accomplishments in those heady days. As the year wound down, Americans could only stand back and gaze in awe at their progress. "Manhood is magnified from the industrial standpoint," the *Independent* declared in a year-end essay. "1908 has been worth two of 1808, and ten of 1708."

"We said that 1908 was finished, but that is not strictly true," concluded the *Independent*. "It has hardly begun to do its work in history."

A heavy snow fell over Washington a few days before Christmas. Theodore Roosevelt went trudging into the blanketed suburbs with two Secret Service agents in tow, then walked home after dark to a White House filled with family. His sons were all returned, Theodore Jr. back from his job in Connecticut, Kermit from Harvard, Archie from boarding school. Ethel and Quentin were there, too, as were Alice and her husband, Congressman (and future Speaker of the House) Nicholas Long-worth. The stockings were hung for a final time from the mantel over the fireplace. "No doubt some of the things which find their way into those stockings will later find their way into the lonesome wilds of Africa—" speculated the *Washington Post,* "perhaps a handy bowie knife for the President, and a jungle-proof photographic outfit for Kermit."

The Roosevelts aimed to make this last Christmas in the White House a memorable one. They celebrated with an afternoon dinner that was large but intimate. About fifty friends, including a number of children apparently unattached to specific adults (probably friends of the sociable Quentin), joined them at a table decorated with red leaves, ferns, and small gifts wrapped in tissue paper. When Quentin opened his trifle to find a tissue paper crown inside, the table burst out with laughter, then all turned to a Christmas feast of roast turkey, vegetables, and flaming plum pudding, culminating with ice cream molded in the shape of miniature Santa Clauses. After dinner, the children hurried downstairs into the "crypt," a subterranean room lit by lanterns and a glowing Christmas tree. Under the tree a heap of gifts awaited them. Later, Archie Butt would write to his sister-in-law that President Roosevelt "was overflowing with good humor and kissed the youngsters and played with them with the same vim as he talked politics with his other guests."

The joy of Christmas came tinged with sadness, though. Most of the Roosevelt children had spent their childhoods in this house they would soon be leaving. Quentin, the youngest, had been only three when the family moved in; his earliest memories were of running through the corridors and gardens of the executive mansion. Even Alice, who was already a teenaged girl when she moved in with her father and had become a young woman amid the glitter of the White House, strolled though the rooms "unutterably sad," according to Archie Butt, who strolled alongside her.

Archie Butt had his own sad thoughts to bear that first Christmas after his mother's death. Unlike the rest of those in attendance, however, he would not be leaving the White House anytime soon. President-elect Taft had asked him to stay on and he had agreed. The decision would prove fateful. Fractures were already beginning to show in the relationship between past and future administrations. Much to Roosevelt's surprise, Taft had yet to ask any of his cabinet members to remain in their posts past March; indeed, almost no one from Roosevelt's administration would be invited to remain other than Archie Butt. Meanwhile, Mrs. Taft was announcing changes in the way social functions would be handled at the White House. She had decided, for instance, to dispense with the frock-coated White House ushers and replace them with liveried footmen, an implicit criticism of Mrs. Roosevelt's more formal tastes.

These were small seedlings of discontent, but they would soon grow into intractable grudges. Four years from now (who could have imagined this in 1908?) Roosevelt would be back from his trip to Africa so bitterly disappointed in his successor that he would run against him in 1912—not as a Republican but under the banner of the newly formed Progressive party. Archie Butt, affectionately faithful to both men, would suffer the break between Roosevelt and Taft harder than anyone. His health racked by the emotional stress of divided loyalties, he'd take a six-week leave of absence from the White House in the spring of 1912 and travel to Europe to recuperate and escape. Returning to America on April 10 of that year, he would board the RMS *Titanic* in Southampton, England, for the ocean liner's maiden voyage. Four days later, the *Titanic* would hit an iceberg in the North Atlantic, and Archie Butt, along with more than fifteen hundred others, would drown.

And, of course, so much more was to come. Two years after Archie Butt's death, World War I would break out in the Balkans. A year after that, in May of 1915, German submarines would sink the *Lusitania*, killing 1,198 aboard and rousing many Americans, including former President Theodore Roosevelt, to press for war. When America finally declared against Germany in 1917, Roosevelt's four sons would hurry to the front to serve. The youngest, Quentin—his father's favorite, the boy who lightened his days with mischief and baseball—would leave Harvard to join the newly formed Air Service of the U.S. Army. The three oldest boys would all be wounded. Quentin would be killed while flying over France in the summer of 1918, shot down by a German plane. His father, heartbroken, would die six months later.

The future was marvelous and ghastly. Better to enjoy Christmas without knowing what was coming. Like the dogs Frederick Cook took with him to the Pole, who yapped joyfully the morning they left Annoa-

The Roosevelt family outside the White House on Christmas Day, 1908.
From left to right: Ethel, Kermit, Quentin, Mrs. Roosevelt (Edith),
Theodore Jr., President Roosevelt, Archie, Alice, and Congressman
Nicholas Longworth (Alice's husband and the president's son-in-law).

tok. Or like children who believed the North Pole was the home of
Santa Claus. Better not to know, just yet, the cold truth.

True North

A few hundred miles shy of the *real* North Pole, Robert Peary and the
crew of the real *Roosevelt*—not the fictional ship sunk by Germans in
H. G. Wells's novel—celebrated Christmas with a day of feasting and
tournaments. The winter solstice had fallen a few days earlier and darkness was constant along the southern shore of the Arctic Ocean. Nature
cooperated on Christmas by providing a relatively warm day of minus
twenty-three Fahrenheit and a festive display of boreal lights.

The crew members found ways to entertain themselves. They laid out
a racecourse on the ice, using two long rows of ships' lanterns to light
it—"a strange sight," Peary wrote, "seven and a half degrees of the earth's
end." On this illuminated runway they raced for prizes, Americans and
Eskimos, men and women alike. Other contests included tug-of-war,

wrestling, and dice-throwing. All of this was topped off by a dinner of musk ox, plum pudding, and sponge cake.

Frederick Cook passed a less festive Christmas a few degrees farther to the south. He huddled in the cold cave he shared with his two Eskimo companions, Etukishook and Ahwelah. Two candles of musk-ox fat provided a feeble light. In the gloom of the cave the men could hardly see each other's faces. There was nothing to do but sit and wait for the darkness to pass.

The sky had been dark since early November. The men did not go out much; fear of bears kept them inside. They did the best they could to make their cramped domicile habitable, cladding the walls with pelts and making beds of furs, but the most that could be said of it was that it beat freezing to death or getting devoured by bears. The days of endless nights passed slowly and uncomfortably. Twice a day they ate a joyless meal of boiled meat and tallow. In the remaining hours, Frederick Cook's mind wandered back over his life, recalling forgotten friends and incidents that now seeped from the recesses of his brain into consciousness.

> The hopes of my childhood and the discouragements of my youth filled me with emotion; feelings of pleasure and sadness came as each little thought picture took definite shape; it seemed hardly possible that so many things, potent for good and bad, could have been done in so few years. I saw myself, not as a voluntary being, but rather as a resistless atom, predestined in its course, being carried on by inexorable fate.

Inexorable fate would not be kind to Cook in the months and years ahead. After Christmas and the New Year, in the first glimmers of February light, he and his companions would begin a harrowing two-month journey back to Annoatok. The three men, reduced to eating pieces of leather to fend off starvation, would arrive on the shore of Greenland as gaunt creatures back from the dead. Following another long and difficult sledge journey to the Danish settlement of Upernavik, Cook would board a ship to Copenhagen. On September 1, 1909, he would wire news of his triumph to the world: "Reached North Pole April 21, 1908. Discovered land far north."

As luck would have it, just five days later, on September 5, a wire would arrive in New York from Robert Peary telling the world that *he* had discovered the North Pole: STARS AND STRIPES NAILED TO THE

POLE, is how Peary phrased it. Peary claimed he'd reached the North Pole on April 6, 1909, almost a full year after Cook.

As if two men announcing almost simultaneously the conquest of the Pole were not dramatic enough, Peary immediately followed up with another, more startling claim: Frederick Cook had never *gotten* to the Pole; he was a liar, a fraud, who had "simply handed the public a gold brick."

It was unclear at first upon what, other than his own intuition, Peary based his accusation. In time, though, evidence against Cook was produced. Peary's assistants had personally interviewed Etukishook and Ahwelah and the two Eskimos had admitted (according to Peary) that they were never more than two day's hike from land. Cook's account of his journey, moreover, was riddled with inconsistencies. His reports of solar effects, for example, did not match the corresponding latitude he claimed to have reached. The amount of food he took could not have lasted the number of days he claimed to be on the ice, even allowing for the consumption of dog meat along the way.

And then there was that mysterious discovery of Bradley Land, which Cook described so vividly, and which continues to trouble even his defenders to this day. There is no land where Cook sighted it. Possibly what he saw was a giant "ice island," a great heave of ice driven up by the bashing of ice plates. But this is unlikely. The land he described extended fifty miles, making it much larger than any known ice island. So what did he see? Or did he see anything? Did he imagine it? Did he invent it out of thin air, a detail to make the journey sound more plausible? If so, this raises other questions. Surely it would have occurred to him that, in time, others would follow his footsteps to the Pole and find him out. He could easily have furnished a more modest feature of the icescape, one that would have been easier to explain away than an island fifty miles long.

Robert Peary's account had plenty of its own inconsistencies. Peary, too, recorded icescape features where they did not exist, which would seem to put his own claim in doubt. But Peary was tireless in discrediting Cook, and with the weight of his long-standing reputation behind him he made the better case to the press and public. The *New York Times*, having paid Peary four thousand dollars for the rights to his story, had a vested interest in pressing his claim and did so often. The National Geographic Society, too, sided with Peary. Slowly but surely, the preponderance of opinion moved to support the veteran explorer's claim. With the exception of a few loyal defenders, Frederick Cook was hung out to dry in the chill winds of history.

It remains one of the greatest controversies and mysteries in the history of exploration, still unresolved and debated a century later. Did Frederick Cook ever get to the North Pole? And if not, was the whole journey a lie? A fantasy? An epic bout of madness? A dream?

That grueling drive across the Arctic? That fall through the ice? The final feverish dash? The dozens of vivid descriptions of sun-dazzled snow and impenetrable darkness? Did Frederick Cook perpetrate one of the greatest frauds in history, or suffer one of its most unjust attacks?

True or false, Cook's memoir of his journey is a compelling modernist parable. Published several years after his return to New York, *My Attainment of the Pole* is the story of a man who travels beyond the conventions and comforts of civilization and enters an ever-changing world where space and time collapse—as if he'd walked not to the North Pole, but into a diorama of Albert Einstein's Special Theory of Relativity (first published in 1905) or into one of the perspective-scrambling Cubist paintings that Pablo Picasso and Georges Braque were beginning to make in 1908.

Arctic artifice: Frederick Cook poses in a studio after his return home from the north.

Cook's Sisyphean trudge, which is revealed to him as a meaningless act of
ambition, which is then dismissed as a fraud—this was a tale worthy of
Kafka. If Frederick Cook did not in fact go to the Pole, he deserves a high
place among the modernist fiction writers of the early twentieth century.

You did not have to be a polar explorer, a Cubist painter, or Albert Einstein to appreciate the implications of modernity. Any working-class or
middle-class urban American could grasp the essential topsy-turviness of
modern life; he lived it. Movies rushed him to strange places with their
flickering images of light. Amusement parks turned him upside down and
inside out. Electricity fused his days and nights. Automobiles, if he were
lucky enough to own one, collapsed his sense of space and time, as a two-
day journey shrank to a two-hour journey. It's safe to assume that many
Americans entered 1908 having never seen a movie, ridden in an automobile, or grasped that humans could fly—and ended the year having done
all three. They could not help but come out of 1908 as practicing mod-
ernists, and agreeing with Thomas Edison that *anything, everything, is possible.*

     It was an exhilarating thought, but also a scary one. Where anything
is possible, where the frontier of human experience is ever expanding,
there are no boundaries, no routes, no maps. Unlike the Wild West, such
a world is unconquerable, and therefore daunting. Which may explain
why the future of the twentieth century would be largely a story not of
people embracing new liberties—though there would be those—but of
people submitting to new dogmas and moral imperatives (progressivism,
fascism, communism, fundamentalism) that rescinded liberties in favor
of order and certitude.

The Problem of the Future

From where most Americans sat that Christmas of 1908—which was
almost certainly a better place than a candlelit cave in the Arctic—the
future looked bright. And it would be, too, assuming they survived the
First World War, the Great Depression, and the Second World War, not
to mention the Cold War, Vietnam, and assorted other national and
international traumas, all of which a young man or woman of 1908
might witness over the course of a long life. The standard of living *would*
rise for most Americans. Matters of health and comfort *would* improve,

and not just for the rich but for all Americans. An eight-hour day, pensions, workers' compensation, child labor laws, all but unknown to laborers in 1908, *would* be common benefits in short order, thanks in large part to the efforts of progressives.

Progressives would persist in pressing their moral imperatives on American society, sometimes fairly and compassionately, sometimes piously and coercively. (An example of the latter was provided on Christmas Eve of 1908, when reformers and religious leaders successfully lobbied New York's mayor to temporarily shut down all 550 of the city's nickelodeons, on the ground that both the movies and the darkened theaters encouraged licentiousness.) At their best, progressives did not try to halt the rough progress of the modern world; they tried, rather, to steer it safely and sanely. Their success was mixed, and their fortunes would rise and fall after 1908. In 1912, they would field their own political party—the Progressive party—with Theodore Roosevelt, still "strong as a Bull Moose" (his words), as their candidate. Roosevelt would defeat Taft in the general election but lose to Woodrow Wilson. This would be the beginning of the end for the progressives as a political party.

Broadly speaking, the future of America was going to be progressive anyway. To a degree that would have been inconceivable to most people in 1908, government did take upon itself the burdens of providing for the poor and the sick, while accepting an ever greater role in regulating areas of human concern once considered beyond its purview. Through the eras of Prohibition, the New Deal, the Great Society, and on through Democratic and Republican administrations alike, a strain of progressive thought would infiltrate American politics and influence its course.

One important change with which progressives cannot be credited is the advance of women's rights in America. While some (like Jane Addams) supported suffrage, just as many progressives of 1908 believed a woman's place was in the home, since it was there she could do the most good for the moral and physical health of her children and, by extension, the nation. Suffragettes would eventually win women the right to vote in 1920 with the Nineteenth Amendment. Greater numbers of women would attend college, and more employment opportunities would open to them.

Along with these political and economic changes, many American women would encounter new ideas of themselves as sexual, and not merely domestic or romantic, beings. G. Stanley Hall, the president of

Clark University—and the country's preeminent psychologist (it was he who coined the term "adolescence")—deserves some credit for bringing these ideas to American soil. In mid-December of 1908, he sent a letter to Vienna, Austria, addressed to a middle-aged neurologist named Sigmund Freud. The letter held out an invitation to Freud to visit America and speak of his work at Clark. Freud would accept the invitation and arrive for his first and last visit to the United States in the fall of 1909. Over five consecutive days, he would deliver a series of lectures that would challenge the way Americans thought about sex. He would encourage his American counterparts to cast aside their prim Victorian views of lust as a primitive urge to be expunged from women or men and to think of it, instead, as natural *and* healthy. It was the repression of the sex drive, not its expression, that made people psychologically unbalanced. To *un*repress his patients, including women, Freud proposed to investigate their dreams, those hothouses of the unconscious, and bring their sublimated desires into the light of day.

In retrospect, much of what occurred in America in 1908 resembles one long Freudian dream. Giant phallic towers rise in the sky. Humans soar through the air like mythological creatures. A man walks across a frozen ocean to get to the North Pole—or does he?—while another forgets to touch second base and loses the pennant for Gotham. Riders gallop under moonlit skies singing hymns. A young woman hurries through a metropolis in a revealing dress under thousands of leering gazes. A fleet of white ships sails across the horizon. It didn't take Sigmund Freud to acknowledge what bursting desire and confusion these visions contained.

For the African Americans who had the misfortune to be cast as demons in the fervid imaginations of whites, the reality of 1908 was closer to a nightmare. It would be heartening to report that white Americans discovered their consciences after the Springfield riot in August. But that is not what happened. Race relations did not improve; if anything, they got worse, especially in the north. In 1917, whites would kill thirty-nine blacks in East St. Louis. In 1919, a five-day race war in Chicago would leave twenty-three blacks and fifteen whites dead. Other race riots would occur after World War I in Omaha, Nebraska, and Tulsa, Oklahoma.

Springfield, Illinois, would celebrate the centennial of its most famous father on February 12, 1909. The gala to honor Abraham Lincoln would

be held at Springfield's State Arsenal, the very same building into which
terrified blacks had fled for their lives back in August. This time, the
arsenal would be filled with whites and whites only. No black citizens
would be invited to attend.

Elsewhere, at least, Lincoln's centennial would be observed in the
spirit more in keeping with the legacy of the Great Emancipator. Here
was the silver lining in the dark cloud over Springfield. Galvanized by
the violence of the riot and by the reporting of William English Walling,
sixty prominent men and women, blacks and whites alike, would cosign
a statement of cooperation and intent, a manifesto really, released on
Lincoln's hundredth birthday. The text would read, in part:

> The celebration of the Centennial of the birth of Abraham Lincoln,
> widespread and grateful as it may be, will fail to justify itself if it takes no
> note of and makes no recognition of the colored men and women for
> whom the great Emancipator labored to assure freedom. Besides a day of
> rejoicing, Lincoln's birthday in 1909 should be one of taking stock of the
> nation's progress since 1865. . . . If Lincoln could revisit this country in the
> flesh, he would be disheartened and discouraged.

The attached names were a familiar smattering of progressives and
muckrakers, of college presidents and intellectuals: Jane Addams, Ray
Stannard Baker, Lincoln Steffens, Madeline Doty, W.E.B. DuBois, Ida
Wells-Barnett, John Dewey, and dozens of other prominent black and
white Americans who signed what became the founding document of
the National Association for the Advancement of Colored People. Born
of the sorrow and shame of Springfield, the NAACP would be a beacon
of hope to African Americans throughout the twentieth century.

First, though, came another beacon. On the morning of December 26,
1908, in Sydney, Australia—Boxing Day in the antipodes but still the
waning hours of Christmas in the United States—a thirty-year-old
African American boxer from Galveston, Texas, named Jack Johnson
stepped into the ring to fight Tommy Burns, the heavyweight champion
of the world.

The fight between Jack Johnson and Tommy Burns, a white man of
French-Canadian descent, had been a long time coming. Like every
titleholder before him, Burns had refused to "cross the color line" and

compete against a black man, even a fighter—*especially* a fighter—as obviously gifted as Johnson. But Johnson refused to take no for an answer. He pursued Burns to London, constantly badgering him until even whites began to suspect the Canadian was hiding beneath his white skin. Burns finally agreed to a match in Sydney, but only on condition of a deal that guaranteed him, win or lose, $30,000 of a $35,000 purse. He probably counted on Johnson to balk at the deal, but Johnson took it. He would settle for five thousand dollars for the opportunity to become the first black heavyweight champion of the world.

The fight began at eleven in the morning. Working slowly, deliberately, and effectively, the "big Negro," as the press called him, destroyed Burns before twenty-five thousand spectators. By the fourteenth round, blood was pouring from Burns and police jumped in to stop the fight. The referee declared Johnson the victor.

"Though he beat me, and beat me badly, I still believe I am his master," said Burns after the fight, already calling for a rematch. "I feel the sting of defeat doubly because of the fact that my fall allowed a colored man to usurp the title for the first time in ring history."

Johnson laughed. "Now that the shoe is on the other foot, I just want

Jack Johnson in 1909, shortly after
becoming the first African American
heavyweight champion of the world

to hear that white man come around whining for another chance." Eventually, Burns decided he did not want another chance after all.

Jack Johnson would remain the heavyweight champion of the world for seven years, fending off a series of "Great White Hopes" who tried and failed to take the title back to the Caucasians. He would keep winning, keep badgering and bragging, and run through a series of white girlfriends faster than he drove his cars, which was very fast. In the end, it would be the white women, not the white boxers, who proved Johnson's downfall. He would be sent to jail after federal prosecutors, misapplying a statute meant to discourage prostitution, charged him with illegally transporting a woman across state lines for immoral purposes. His crime had been to send a train ticket to one of his white girlfriends.

That was later, though. Now was Christmas, and Jack Johnson's victory was a late gift for African Americans to savor in the closing moments of 1908.

Fordian Dreams

Of all the forces and tensions that would swirl through America in the years after 1908—political, economic, cultural, military—none would transform the country more thoroughly and rapidly than the internal combustion engine; more specifically, than the millions of internal combustion engines that would be lodged inside Henry Ford's Model Ts. The Model T was not the only car made in America in the years after 1908, nor was it the best or the handsomest. But it was the one that brought the automobile age to the masses and turned America into the first true automobile nation.

Americans at the end of 1908, though aware they had lived through a transformative year in aviation, were mostly oblivious to the fact that a revolutionary automobile had just been hatched in Detroit. In reviewing the year's automotive progress, *Scientific American* confidently reported that "during the past year no improvements of a radical character have been introduced." *Technologically,* perhaps, the Model T was not a radical departure. Commercially, though, it was destined to be the most successful automobile that had ever been manufactured—or ever would be.

The Model T got off to a slow start. Ford Motor manufactured just 309 of them in the three months between its introduction in October of 1908 and the end of the year. Ford was still working out the kinks in production. In these pre-assembly-line days, cars were essentially built in one place, on the third floor of the Piquette Avenue plant. The chassis remained stationary, as men and material came to it. Then each completed automobile was loaded onto the freight elevator and lowered to the ground for shipping.

Even using old-fashioned methods, though, Ford and company quickly learned to supply cars faster. In 1909, when Ford suspended production of all other models to focus entirely on the T (and uttered his oft-quoted maxim, "The customer can have any color he wants, so long as it's black"), the company produced more than fourteen thousand Model Ts. By 1911, after moving into the enormous Highland Park plant, the company was producing fifty thousand a year.

The Model T proved just how eagerly America had been awaiting such a machine. Given its vast geographic distances, it was almost as if the country had been made for the automobile, rather than the other way around. The railway had already tracked much of America, but rail travel was limited by station stops and train schedules. Automobiles, by contrast, allowed Americans to get up and go, whenever, wherever they pleased. A journey that would have taken all day on horseback could be accomplished in less than an hour. For the 50 percent of Americans who still lived rural existences, the automobile was an instant solution to the isolation of country life. They could go into town more easily to purchase goods, travel across the state to inspect livestock, or simply make social calls on distant neighbors. With the possible exception of the telephone, the Model T was the most sociable machine ever invented.

At the same time that rural Americans discovered its pleasures, urban Americans, who needed *less* human contact, would find in the Model T an escape from congestion. With a press of the pedal, they could be out in the countryside in no time, the clean revivifying air rushing through their lungs and putting a glow in their cheeks. Not only might they visit the countryside on a whim, they might choose to *move* there, while keeping employment in the city. Suburban communities had already sprung up along trolley tracks running from the cities, but the automobile allowed its owners to live virtually anywhere. The country mouse and the city mouse could be one, just so long as the mouse had a Model T.

Because the Model T was easy to operate and dependable, women could feel confident at its wheel and enjoy freedom of movement as never before. (Ford understood this and advertised the car as a vehicle of women's liberation.) In a nation populated by millions of restless souls, the car was an itinerant's best friend. For a nation of sexually charged young men (and, yes, young women), the backseat of the Model T afforded a snug private space where couples might find precious privacy. Automobiles transported people in more ways than one.

As the Model T changed how people lived, it also changed, in time, how they worked. In 1913, Ford would institute the assembly line at his Highland Park plant. Here the manufacturing of a Model T was broken down into hundreds of discrete tasks. As the chassis of the automobile moved automatically along a track, passing through a gauntlet of men and parts, each man would repeat one dedicated action. One worker placed a part, the next placed the screw to hold it, the next tightened the nut of the screw—and so on down the line, men functioning more like machines than men.

The Ford assembly line would be the manufacturing marvel of its age. In the line's first year in operation, the company more than doubled its output of Model Ts from 82,000 to 189,000, about half the automobiles manufactured in America that year. By 1916, Ford would be manufacturing almost six hundred thousand cars a year and could lower the price of the Model T to $360, which produced more demand, to which Ford responded with more supply.

Assembly line production was a good news/bad news proposition for the American proletariat. While allowing manufacturers like Ford to drop prices so that even an assembly-line worker could afford one, it also subjected those workers to day after day of repetitive labor. Their power and freedom as consumers in their nonworking hours—automobiles, entertainment, mounds of inexpensive savory food—would be purchased by hours of mindless toil. This was, in many ways, a Faustian bargain. But it was also a deal most Americans would be unable to refuse.

One of the immediate effects of Ford's assembly line was that it drove smaller automobile manufacturers out of business. A few large competitors, like General Motors (founded in 1908), grew alongside Ford, but the vast majority went under. The Thomas Motor Company was among the victims. All the publicity from the race around the world had generated a temporary boost in sales, but this was not enough to sustain the company. Thomas went bankrupt in 1913, the same year Ford launched his assembly line.

It is impossible to quantify all the myriad effects, good or bad,

impressed on American life by the automobile after 1908. Today, the automobile's reputation is tainted by developments Americans of 1908 never
anticipated. The forty thousand Americans killed in accidents every
year. The suburban culs-de-sac gobbling up countryside. The strip malls
and superstores where small towns wither. The hours of life and liberty
lost in gridlock. The millions of gasoline-guzzling internal combustion
engines polluting the air, warming the earth.

But all of these were, in 1908, very much problems of the future.

What the Sky Held

Two days after Christmas, early on the morning of December 27, acolytes
of a self-proclaimed prophet named Lee J. Spangler met in a snow-
covered graveyard on a hilltop in Nyack, New York, a Hudson River town
north of New York City. The women wore white dresses made specially
for the occasion. That occasion, Spangler had assured them, was Judgment
Day. They waited all day in the graveyard for the world to end. When it
did not, they dispersed and went home.

Later that night—already the following day on the European continent, the early hours of December 28—the world did come to an end. A
massive earthquake, followed immediately by a large tsunami, leveled a
wide swath of southern Italy. Centered in the Strait of Messina, between
Sicily and the Italian mainland, the quake was the strongest in modern
Europe's history. The entire city of Messina crumbled to the ground, and
almost its entire population perished.

For Americans, who responded at once by sending money and aid to
Italy, the scale of the disaster was inconceivable. The terrible San Francisco earthquake of 1906 had killed three thousand people. The toll in
Italy topped two hundred thousand. No act of man or nature would
equal this kind of devastation until America dropped atomic bombs
from airplanes on Nagasaki and Hiroshima in 1945.

On December 31, 1908, the last day of the year, Wilbur Wright emerged
from his shed at Camp d'Auvours and ducked into his Flyer. The day
was cold and gray in central France. Snow dusted the ground. Despite
the weather and the depressing tragedy unfolding in southern Italy, a
large and giddy crowd was present. People came to see Wilbur Wright
perform one last extraordinary flight before the year was done. Today,

Wilbur intended to win the Coupe de Michelin, worth twenty thousand francs, or about four thousand dollars. He planned to fly longer than either he or Orville—longer than anyone—had ever flown.

The weather was poor for flying, only slightly better than the previous day, when Wilbur had been forced down after the oil in his engine froze and his wings iced over. Now, shortly after he took off, sleet and rain began to fall. At an altitude of two or three hundred feet, traveling at about forty miles per hour, the cold was bitter. Wilbur would have preferred to be back at the overheated house in Dayton for the New Year, rather than freezing over Le Mans, but he was determined to establish one last record for the Wrights. "If I had gone away the other fellows would have fairly busted themselves to surpass any record I left," he wrote to his father. "The fact that they knew I was ready to beat anything they should do kept them discouraged."

He could not go home to his family, but, as he now knew, his family was coming to him. In a few days, Orville and Katharine would be boarding a ship in New York and sailing to Europe. The three siblings would then travel to the south of France, where they would establish a winter camp, and where Wilbur, as part of his contract with the French syndicate, would train three pilots to fly a Wright plane—the first flying school in history.

It was not like Wilbur to let his mind wander while flying in such challenging conditions, but at some point that day his thoughts must have turned, if only fleetingly, to the year just passed. The year 1908 had been astonishing for the Wrights—and astonishing for the world in large part *because* of them. When it began, Wilbur and Orville were widely considered frauds; and those who didn't think they were frauds had probably never heard of them. The idea that they, or anyone, might achieve sustained flight for *minutes* would have seemed far-fetched to most people. Anyone proposing to fly for *hours* might have found himself joining Harry Thaw in Matteawan.

Since January 1, 1908, Wilbur had nearly died at Kitty Hawk; Orville had come closer to death at Fort Myer; Thomas Selfridge *had* died. The Wrights had become heroes. Theodore Roosevelt, who now shared his stature as the most celebrated American on earth with the two brothers from Dayton, had just announced his hope to pin a medal on them before he left office.

How much pleasure Wilbur got out of all this is unclear. Mostly, he worked and fretted and flew. Celebrity did not enthrall him. Flying was more pressure than pleasure; after this flight on the last day of 1908, he would take to the air only a few more times in his life (most notably in New York City during the Hudson Fulton celebration in early October of 1909). Once he was able to stop, he did.

His attention and time after 1908 would be increasingly consumed by troublesome legal battles with the Wrights' rivals, mainly Glenn Hammond Curtiss, regarding patents. That time was short, in any case. Exactly four years after he arrived in France to astonish the world, Wilbur would die, on May 30, 1912, at the age of forty-five. It would not be flying that killed him but typhoid fever.

Perhaps on that last day of the year, soaring through the cold wind and sleet over Le Mans, Wilbur allowed himself at least a moment of pure joy in savoring what he and Orville had achieved. Though he never articulated it, plenty of others did. "In tracing the development of aeronautics, the historian of the future will point to the year 1908 as that in which the problem of mechanical flight was first mastered," *Scientific American* would state in its first issue of the New Year; "and it must always be a matter of patriotic pride to know that it was two typical American inventors who gave to the world its first practical flying machine."

Wilbur flew for one hour, and then for another, and still kept flying. Finally, after two hours and twenty minutes, he landed. He had demolished every endurance record of the last year and had flown a distance of nearly ninety miles. Had he been flying a straight line in the direction of Paris, rather than around and around in circles, he could have touched down in the gardens at Versailles.

That evening, Wilbur celebrated the arrival of the New Year in Le Mans at a reception in his honor. Then he returned to Camp d'Auvours and went to sleep in his shed.

As the clock struck midnight in France, New Year's Eve celebrations were just getting under way in New York. The restaurants were booked with overflow crowds, even greater than last year's. The night was cold and people sought the warmth of indoors. The Broadway theaters, like the restaurants, were filled to capacity. By midnight, though, the theaters had let out and all those capacity crowds had poured onto Broadway to join the shivering mass already packed into Times Square, gazing up at the bright ball that hovered over the Times building.

A few days earlier, a writer named Edwin Wildman had climbed up to the top of the building and looked down at Times Square, imagining the crowds and the noise that would fill it on New Year's Eve. In an essay in the *Times*, written in a style at once archly old-fashioned yet hallucinatory enough to suggest that its author, had he been born a few generations later, might have enjoyed the 1960s, Mr. Wildman pondered the past and the future. "One hundred years from tonight no living soul who treads that great White Way will draw from eyries of memory the story of Now," he wrote. "Where are the men of yesterday—the yesterday of one hundred years? Gone, too, as will the men of to-day that century hence." Wildman continued:

I look down upon the vast throng of surging men and women crying the welcome of the New Born Year and think of the to-morrows. I gaze below through the mist and see the twinkle of the upturned Milky Way; I hear the clash of discordant sounds, the murmur of multitudinous voices, the roar of a human symphony, and I know it is mind that is speaking—the oneness of a vast intelligence moving on: the echo of a single thought: the vibration of a single pulse.

That pulse, Wildman seemed to hope, would keep beating long after his heart and the hearts of the thousands below had stopped.

It was almost midnight. The bright light shone way up at the top of the Times building. The crowd below began to count backward, waiting for the electric ball to descend and the future to begin.

ACKNOWLEDGMENTS

AND SOURCES

Like many of the episodes it describes, this book was something of a quixotic and fabulous adventure from the start. I am grateful to many people for taking a leap of faith with me, then helping me keep the faith along the way. I owe much thanks to Kris Dahl, my peerless agent, for her sound advice. Thanks also to Sarah McGrath, for her early enthusiasm, and to the wonderful Nan Graham at Scribner.

It was this book's great fortune to end up in the hands of Colin Harrison, its editor. While any faults are mine alone, the book is far better for Colin's lucid insights, his ear for language, and his general wisdom. Karen Thompson steered the book agilely and kindly through the editing process.

The knowledgeable librarians at the New York Historical Society, where I spent many happy days, were helpful and hospitable. (Knowing that the Society's building was completed in 1908 made me feel especially welcome there.) The staffs of the New York Public Library, the Library of Congress, the Special Collections and Archives at Wright State University, and the Houghton Library at Harvard were generous with materials and assistance.

I am deeply indebted to my wife, Ann, and to my sons, Willy and Jack—and to all my family and friends—for their patience and understanding throughout the last two years, and for their continuing confidence that I will again, someday, talk about things that have no direct bearing on the year 1908.

To those people to whom I wrote checks accidentally dated '08, my apologies. You will not have to wait long now to cash them.

It may seem odd to thank nameless newspaper reporters who have been dead for many years, but I am obliged to dozens, if not hundreds, of writers who practiced journalism a century ago. This book found its very heart in the newspapers of 1908, especially the *New York Times, New York Herald,* and *New York Tribune*— that great troika—and, to a lesser extent, the *New York World, New York Press, New York Telegraph, Chicago Tribune, Washington Post,* and *Los Angeles Times.* Many other newspapers, metropolitan and regional alike, were consulted. With the exception of Henry Ford's development of the Model T, which occurred behind closed doors, and Frederick Cook's journey to the North Pole, which occurred thousands of miles from the nearest journalist, the events covered in this book were reported as they unfolded. Newspaper writing circa 1908 was hyperbolic, blatantly partial, and frequently purple, but it also spilled over with infectious awe and wonder. I recommend to anyone disposed to this sort of thing (it helps to have a few free weeks and to be slightly mad) to scroll, as I did, through the full 1908 run of the *New York Times* on microfilm.

Along with newspapers, weekly and monthly journalism added a great deal to my appreciation and comprehension of America in 1908. Slightly more reflective—and less reflexive—than the daily feed, magazines like *Harper's Weekly* (made magically accessible and searchable by HarpWeek) provided a good rehash week by week of the news of the day. *Hampton's* (later renamed *Hampton's Broadway*) was a happy revelation, a magazine with more personality and far-sightedness than most. I also leaned heavily on *Scientific American* and *Collier's*.

Beyond the contemporary journalism, primary sources for this book include published letters, journals, and memoirs, especially the works of Frederick Cook, the Wrights, and Archie Butt. Several recently published books, too, deserve special mention. Michael McGerr's *A Fierce Discontent* is a terrific overview of progressivism. Julie Fenster's *Race of the Century* fleshes out the contours of the New York–to–Paris race, and Bruce Henderson's opinionated but even-handed *True North* helped me piece together the strange trips of Frederick Cook and Robert Peary. James Tobin's *To Conquer the Air* was very useful (and entertaining) regarding the Wrights. Two recent biographies of Henry Ford, *Wheels for the World* by Douglas Brinkley and *The People's Tycoon* by Steven Watts, give comprehensive views of Ford and his Model T. Last, no two volumes provide more insight into Theodore Roosevelt than Edmund Morris's delightful *The Rise of Theodore Roosevelt* and *Theodore Rex*.

What follows is a list of selected sources. Newspaper and magazines have generally been cited within the text of the book, but where a particular periodical is relied on extensively I cite it again here.

SOURCES BY SUBJECT

African Americans and the Springfield Riot

Baker, Ray Stannard, *Following the Color Line* (1908; 1964).

Chicago Daily Tribune. Extensive coverage of Springfield riot. August 15, 1908–August 20, 1908.

Crouthamel, James L., "The Springfield Race Riot of 1908," *The Journal of Negro History*, July 1960.

"Mob Rule in Lincoln's Town," *Harper's Weekly*, August 29, 1908.

Ovington, Mary White, "How NAACP Began," NAACP website (www.naacp.org/about/history).

Senechal, Roberta, *The Sociogenesis of a Race Riot: Springfield, Illinois, in 1908* (1990).
Walling, William English, "The Race War in the North," *The Independent*, September 3, 1908.

Anarchists and Night Riders

Adamic, Louis, *Dynamite* (1934).
Goldman, Emma, "What I Believe," *New York World*, July 19, 1908.
Miller, John G., *The Black Patch War* (1936).
Nall, James O., *The Tobacco Night Riders of Kentucky and Tennessee* (1939).
"The Night Riders: More Desperate Than the Ku Klux Klan," *New York Herald*, January 19, 1908.
Roth, Walter, and Joe Kraus, *An Accidental Anarchist: How the Killing of a Humble Jewish Immigrant by Chicago's Chief of Police Exposed the Conflict between Law & Order and Civil Rights in Early 20th-Century America* (1998).

Automobiles and the Model T

Brinkley, Douglas, *Wheels for the World: Henry Ford, His Company, and a Century of Progress* (2003).
Clymer, Floyd, *Henry's Wonderful Model T, 1908–1927* (1955).
Fitch, George, "The Automobile," *Collier's*, September 19, 1908.
Flink, James J., *America Adopts the Automobile, 1895–1910* (1970).
Ford, Henry (with Samuel Crowther), *My Life and Work* (1922).
Sorensen, Charles E., *My Forty Years with Ford* (1956).
Stern, Philip Van Doren, *Tin Lizzie: The Story of the Fabulous Model T* (1955).
Watts, Steven, *The People's Tycoon: Henry Ford and the American Century* (2005).

Automobile Race, New York to Paris

Cole, Dermot, *Hard Driving: The 1908 Auto Race from New York to Paris* (1991).
Fenster, Julie, *Race of the Century: The Heroic True Story of the 1908 New York to Paris Auto Race* (2005).
McConnell, Curt, *Coast-to-Coast Auto Races of the Early 1900's: Three Contests that Changed the World* (2000).
"The New York to Paris Automobile Race—1908," *Buffalo Tales* (bchs.kearney.net/btales), July and August 1993.
New York Times. Extensive coverage, January 1908–August 1908.
Villard, Henry Serrano, *The Great Road Races, 1894–1914* (1972).

Aviation and the Wright Brothers

Combs, Harry (with Martin Caidin), *Kill Devil Hill: Discovering the Secret of the Wright Brothers, 1899–1909* (1979).
Crouch, Tom D., *The Bishop's Boys: A Life of Wilbur and Orville Wright* (1989).

New York Herald. Extensive coverage, May and September 1908.

Tobin, James, *To Conquer the Air: The Wright Brothers and the Great Race for Flight* (2003).

Washington Post. Extensive coverage, September 1908.

Wells, H. G., *The War in the Air: And Particularly How Mr. Bert Smallways Fared While It Lasted* (1908).

Wohl, Robert, *A Passion for Wings: Aviation and the Western Imagination, 1908–1918* (1994).

Wright, Wilbur, and Orville Wright (ed. Fred C. Kelly), *Miracle at Kitty Hawk: The Letters of Wilbur and Orville Wright* (1951).

Wright, Wilbur, and Orville Wright (ed. Marvin W. McFarland), *The Papers of Wilbur and Orville Wright* (1953).

Baseball

Anderson, David W., *More Than Merkle: A History of the Best and Most Exciting Baseball Season in Human History* (2000).

Deford, Frank, *The Old Ball Game: How John McGraw, Christy Mathewson, and the New York Giants Created Modern Baseball* (2005).

Fleming, G. H., *The Unforgettable Season: The Cubs, Giants, and Pirates, and the Greatest Race of All Time* (1981).

New York Herald. Extensive coverage, September 23–October 9, 1908.

New York Times. Extensive coverage, September 23–October 9, 1908.

New York Tribune. Extensive coverage, September 23–October 9, 1908.

Thornely, Stew, *Land of the Giants: New York's Polo Grounds* (2000).

Great White Fleet

Evans, Robley, "Admiral Evans' Own Story of the American Navy," *Hampton's,* October 1908.

Harper's Weekly:

"Our Armada: Its Precursors and Significance," January 4, 1908.

"The First Page of the Battle Fleet's Log," January, 18, 1908.

"American Navy Second in the World," January 18, 1908.

"The Truth About the Navy," February 8, 1908.

"'Crossing the Line' with the Fleet," February 22, 1908.

"Leaves from the Log," February 29, 1908.

"Through the Straits," April 18, 1908.

Jones, Robert D. (ed.), *With the American Fleet from the Atlantic to the Pacific* (1908).

Matthews, Franklin, *With the Battle Fleet* (1908).

Miller, Roman J., *Pictorial Log of the Battle Fleet Cruise Around the World* (1909).

"The Needs of our Navy," *McClure's,* January 1908.

Reckner, James R., *Teddy Roosevelt's Great White Fleet* (1988).

Wimmel, Kenneth, *Theodore Roosevelt and the Great White Fleet: American Sea Power Comes of Age* (1998).

Nature

Nash, Roderick, *Wilderness and the American Mind* (1967).
Roosevelt, Theodore, *Address of President Roosevelt at the Opening of the Conference on the Conservation of Natural Resources, at the White House Wednesday morning, May 13, 1908, at 10:30 o'clock*, Government Printing Office (1908).

New York

Howe, Wirt, *New York at the Turn of the Century, 1899–1916* (1946).
King, Moses, *King's Views of New York, 1908–1909* (1909).
Poole, Ernest, "Cowboys of the Sky," *Everybody's Magazine*, November 1908.
Van Dyke, John C., *The New New York: A Commentary on the Places and the People* (1909).
Whibley, Charles, *American Sketches* (1908).

The New York Times

Berger, Meyer, *The Story of* The New York Times: *1851–1951* (1951; 1971).
Davis, Elmer Holmes, *History of the* New York Times, *1851–1921* (1921; 1971).
Johnson, Gerald W., *An Honorable Titan: A Biographical Study of Adolph S. Ochs* (1946).
Jones, Alex S., and Susan E. Tifft, *The Trust: The Private and Powerful Family Behind the* New York Times (1999).
"The *Times* Ten Years at One Cent," *New York Times*, October 10, 1908.

North Pole

Cook, Frederick A., *My Attainment of the Pole* (1911).
Henderson, Bruce, *True North: Peary, Cook, and the Race to the Pole* (2005).
Herbert, Wally, *The Noose of Laurels: Robert E. Peary and the Race for the North Pole* (1989).
Hunt, William R., *To Stand at the Pole: The Dr. Cook–Admiral Peary North Pole Controversy* (1981).
Peary, Robert E., *The North Pole: Its Discovery in 1909 Under the Auspices of the Peary Arctic Club* (1910).
Shackleton, Ernest, *The Heart of the Antarctic: Being the Story of the British Antarctic Expedition, 1907–1909* (1909; 1999).
Weems, John Edward, "Peary or Cook: Who Discovered the North Pole?" *American Heritage Magazine*, April 1962.

Popular Entertainment

Batchelor, Bob, *The 1900s: American Popular Culture Through History* (2002).
"Mechanical Joys of Coney Island," *Scientific American*, August 15, 1908.

Peiss, Kathy, *Cheap Amusements: Working Women and Leisure in Turn-of-the-Century New York* (1986).
Singer, Ben, "Manhattan Nickelodeons: New Data on Audiences and Exhibitors," *Cinema Journal,* Spring 1995.

Progressivism

Croly, Herbert, *The Promise of American Life* (1909).
Davis, Allen F., *American Heroine: The Life and Legend of Jane Addams* (1973).
Diner, Steven J., *A Very Different Age: Americans of the Progressive Era* (1998).
Hofstadter, Richard, *The Age of Reform* (1955).
Levy, David W., *Herbert Croly of the New Republic: The Life and Thought of an American Progressive* (1985).
MacDonald, Arthur, *Juvenile Crime and Reformation, Including Stigmata of Degeneration. Being a Hearing on the Bill (H.R. 16733) to Establish Laboratory for the Study of Criminal and Defective Classes* (1908).
Macleod, David I., *The Age of the Child: Children in America, 1890–1912* (1998).
McGerr, Michael, *A Fierce Discontent: The Rise and Fall of the Progressive Movement in America* (2003).
Travis, Thomas, *The Young Malefactor: A Study in Juvenile Delinquency, Its Causes and Treatment* (1908).
Wiebe, Robert H., *The Search for Order: 1877–1920* (1967).

Rich Americans

"American Wives and Foreign Husbands—By an Ex-Diplomat," *Saturday Evening Post,* January 4, 1908.
Brough, James, *Consuelo: Portrait of an American Heiress* (1979).
Chernow, Ron, *The House of Morgan: An American Banking Dynasty and the Rise of Modern Finance* (1990).
Forbes, John Douglas, *J.P. Morgan, Jr., 1867–1943* (1981).
Hovey, Carl, *The Life Story of J. Pierpont Morgan: A Biography* (1911).
Sinclair, Andrew, *Corsair: The Life of J. Pierpont Morgan* (1981).
Sinclair, Upton, *The Metropolis* (1908).
Strouse, Jean, *Morgan: American Financier* (1999).

Roosevelt

Brands, H.W., *T.R.: The Last Romantic* (1997).
Butt, Archie, *The Letters of Archie Butt, Personal Aide to President Roosevelt* (1924).
Hale, William Bayard, *A Week in the White House with Theodore Roosevelt: A Study of the President at the Nation's Business* (1908).
Morris, Edmund, *The Rise of Theodore Roosevelt* (1979).
Morris, Edmund, *Theodore Rex* (2001).
Roosevelt, Theodore, *An Autobiography* (1913).

Roosevelt, Theodore (ed. H. W. Brands), *The Selected Letters of Theodore Roosevelt* (2001).

Roosevelt, Theodore (ed. Joan Paterson Kerr), *A Bully Father: Theodore Roosevelt's Letters to His Children* (1995).

Psychology and Sexuality

D'Emilio, John D., and Estelle B. Freedman, *Intimate Matters: A History of Sexuality in America* (1988; 1997).

Drake, Emma F. Angell, M.D., *What a Young Wife Ought to Know* (1901).

Hale, Nathan G., *Freud and the Americans: The Beginnings of Psychoanalysis in the United States, 1876–1917* (1971).

Hall, G. Stanley, *Adolescence: Its Psychology and Its Relation to Physiology, Anthropology, Sociology, Sex, Crime, Religion and Education* (1904).

Kennan, George, "The Problem of Suicide," *McClure's*, June 1908.

Schneider, Dorothy, and Carl J. Schneider, *American Women in the Progressive Era, 1900–1920* (1993).

Thaw

Langford, Gerald, *The Murder of Stanford White* (1962).

Lutes, Jean Marie, "Sob Sisterhood Revisited," *American Literary History* (2003).

Mackenzie, F.A. (ed.), *The Trial of Harry Thaw* (1928).

Thaw, Mary Copley, *The Secret Unveiled: A Pamphlet* (1909).

ADDITIONAL SOURCES

Adams, Henry, *The Education of Henry Adams* (1907).

Allen, Frederick Lewis, *The Big Change: America Transforms Itself, 1900–1950* (1952).

Andrist, Ralph K., *The Long Death: The Last Days of the Plains Indians* (1964; 2001).

Byington, Margaret, *Homestead: The Households of a Milltown* (1910; 1974).

Cable, Mary, *American Manners and Morals: A Picture History of How We Behaved and Misbehaved* (1969).

Evans, Harold, *They Made America: From the Steam Engine to the Search Engine: Two Centuries of Innovators* (2004).

Fischer, Claude S., *America Calling: A Social History of the Telephone to 1940* (1992).

Johnson, Paul H., *A History of the American People* (1997).

Lord, Walter, *The Good Years: From 1900 to the First World War* (1960).

New York State Census for New York County, New York, 1905.

Novick, Sheldon M., *Honorable Justice: The Life of Oliver Wendell Holmes* (1989).

Shifflett, Crandall, *Almanacs of American Life: Victorian America, 1876–1913* (1996).

Smith, Page, *America Enters the World: A People's History of the Progressive Era and World War I* (1985).

Sullivan, Mark, *Our Times: The United States, 1900–1925,* Volume I (1926).

Tompkins, Vincent, *American Decades, 1900–1909* (1996).

U.S. Census Bureau, *Selected Historical Decennial Census Population and Housing Counts,* Census Bureau website (www.census.gov).

World Almanac and Encyclopedia (1908–1911).

Zinn, Howard, *A People's History of the United States, 1492–Present* (1980).

PHOTOGRAPH CREDITS

INDEX

Page numbers in *italics* refer to illustrations.

Woolworth Building, 118
Working class, 4, 11, 25, 148, 176, 220, 270
World War I, 2, 236, 257, 265, 270
World War II, 258, 270, 278
Wright, Katharine, 122, 168, 190, 204, 279
Wright, Orville, 1–2, 19, 25, 111, 121–23, *123*,
    124–39, 167–73, 185–206, 220, 235,
    254, 255, 278–80
  army contract, 128, 188–89, 193, 198–99,
    205
  character of, 122–24, 189–90
  fame of, 197
  Fort Myer crash, 202–204, *203–204*,
    255, 279
  Fort Myer flight trials, 1, 2, 187–95, *195*,
    196–204, *201–203*, 254, 255, 279
  Huffman's Prairie trials, 126, 130, 136,
    193
  Kill Devil Hill flights, 121–24, 129–32,
    136–39
  Kitty Hawk success, 1, 126
  press on, 130–31, *131*, 132, 136–37, *137*,
    138–39, 191–205, *201–203*, 256
  rivals of, 124–25, 138–39, 167, 169, 188,
    189, 193, 204, 280
  technology, 126, 127, 129–30
Wright, Wilbur, 1–2, 19, 25, 111, 121–23, *123*,
    124–39, 167–73, 185, 220, 235, 253–56,
    278–80
  army contract, 128, 188

  character of, 122–24, 186–87, 189–90
  death of, 280
  fame of, 197, 253–55, 279–80
  Huffman's Prairie trials, 126, 130, 136,
    193
  Kill Devil Hill flights, 121–24, 129–32,
    136–39
  Kitty Hawk success, 1, 126, 279
  Le Mans flights, 1, 128, 167–72, *172*, 173,
    186–87, 191, 197, 200–201, 205–206,
    206, 254–55, 278–80
  Orville's Fort Myer crash and,
    204–205
  press on, 130–31, *131*, 132, 136–37, *137*,
    138–39, 168–69, 171, 191, 205, *206*,
    256, 280
  rivals of, 124–25, 138–39, 167, 169, 189,
    204, 280
  technology, 126, 127, 129–30
Wright Flyers, 124, 128, 129–32, 168, 198,
    279
  first public demonstration, 170–71
Wyoming, 60

Yokohama, 149, 236, 242, 243, 244

Zangwill, Israel, 6
  The Melting Pot, 235, 237
Zimm, Bruno, 92
Zust, 68, 83, 84, 150

Turn the page to read an excerpt from Jim Rasenberger's new book,

The Brilliant Disaster: JFK, Castro, and America's Doomed Invasion of Cuba's Bay of Pigs

1

"Viva Castro!"

April 1959

★

WHEN, REALLY, DID it begin? Was it the day in 1956 Fidel Castro landed on the coast of Cuba with a small band of followers to begin his quixotic campaign against the dictator Fulgencio Batista? Was it in the early-morning hours of January 1, 1959, when Batista, dressed in his New Year's Eve tuxedo, piled cash and family members into an airplane and fled to the Dominican Republic, leaving the country in Castro's hands? Was it, rather, fourteen months later, March 17, 1960, the day President Eisenhower approved the CIA's "Program of Covert Action" to unseat Castro?

Or did the trouble between Cuba and the United States reach much deeper into the past, to Teddy Roosevelt's triumph at San Juan Hill, to the U.S. victory in the Spanish-American War and the succeeding decades of American intervention in Cuban affairs?

In fact, all of these episodes, and many more, were stations on the way to the U.S.-backed invasion of Cuba. Easier than narrowing the origins of the conflict is identifying the moment when it might have been avoided: when some measure of amity between Cuba and America still seemed possible; when Fidel Castro still appeared to many Americans to be the sort of man they could live with, accommodate, and even, per-haps, admire.

Such a moment arrived two years before the first bombs fell on Cuba—two years, in fact, *to the day*. Dwight Eisenhower was beginning his seventh year as president. Richard Nixon was vice president and the presumptive Republican nominee for the 1960 presidential election. John Kennedy was the junior senator from Massachusetts, still mulling a run at the Democratic presidential nomination. And Fidel Castro, three and a half months after conquering Cuba, was celebrating his triumph with a visit to America. His entry into U.S. airspace was a kind of invasion in its own right—a charm offensive. He came bearing a hundred cases of rum, countless boxes of Cuban cigars, and a warm *abrazo* for every man, woman, and child he met.

Castro landed on the evening of April 15, 1959, at Washington, D.C.'s National Airport. The time was two minutes after nine and Castro, typically, was two hours behind schedule, but this did nothing to dampen the enthusiasm of the fifteen hundred admirers awaiting him near the tarmac. "Viva Castro!" the crowd erupted as the door of the Cubana Airlines turboprop swung open and the man himself stepped out. "Viva Castro!"

He looked every bit the legendary guerrilla as he stood atop the airplane ramp, bathed in camera lights and the roar of the crowd. He was a large man, more than six feet tall, clad in rumpled green fatigues and high black boots, the uniform he and his cadre had worn through the battles of the Sierra. He carried an army kit on one shoulder, and an empty canvas pistol holster dangled from his belt. And the beard, of course: that famous beard Castro and his fellow *barbudos* had cultivated while fighting Batista's troops in the wilds of Cuba. *Barbudos* translated as "bearded ones" but sounded like "barbarians," a suitable cognate for the "bearded monster," as one U.S. senator had already taken to calling Castro.

"Viva Castro!"

As the shouts rang up from the tarmac, Castro descended, followed by an avalanche of ministers, businessmen, bodyguards, and others whose roles were more difficult to define. The State Department had been trying to get a fix on Castro's entourage for weeks, but the mercurial Cubans kept coming back with new numbers—thirty-five, then seventy, multiplying, at last, to ninety-four. For the State Department, the changing number was just another sign of the chaos that seemed to percolate around Castro wherever he went, as if he were making it up on the spot, dreaming it as it happened—*it* being the revolution, the new Cuba, this maddening creation called Fidel.

Castro was making his first appearance in the United States since riding into Havana at the start of the year, but the thirty-two-year-old rebel leader was no stranger to America. Newspapers and magazines had been tracking his exploits for years. The general outlines of his biography were familiar: the privileged but combative boyhood as the illegitimate son of a wealthy landowner in eastern Cuba; the student years at Havana University, where he earned both a law degree (hence the honorific "Doctor" often attached to his name) and, more important, a devoted following in the bloody *gangsterismo* political scene of 1940s Havana. There had been a short-lived marriage in 1948, when he took a break from school and politics to travel with his new bride to America for an extended honeymoon. Given what came later, the most remarkable fact about this sojourn was how comfortably Castro had fit into the belly of the imperialist beast, studying English in New York City and enjoying the fruits of capitalism (including a new Lincoln) as much as any other red-blooded young man in postwar New York. Nonetheless, he had returned to Cuba after three months and resumed his life as a revolutionary.

It was in the summer of 1953 that Castro had first come to the attention of the American press. On July 26 of that year he led an attack on an army barracks in Santiago de Cuba. The attack failed miserably but made Castro a hero in Cuba. Castro was sentenced to fifteen years in prison. Batista, in a gesture of goodwill unbecoming of a ruthless dictator, released him after just two. Castro went into exile in Mexico City, befriended a young Argentinean doctor named Ernesto "Che" Guevara, and raised funds to support a new attack on Cuba. In December 1956, in the company of Guevara, his brother Raúl Castro, and seventy-nine other men, Castro sailed across the Gulf of Mexico and through the Yucatán Strait to mount another quixotic attack on Batista's forces. Nearly all the rebels were killed within days of landing, including Fidel Castro, according to Batista's government. And so the world believed until Herbert Matthews, a writer for the *New York Times*, managed to track Castro down in the Sierra Maestra, where he found the rebel leader hiding among the peasants like a modern-day Robin Hood, not only alive but apparently prospering and gathering forces. Matthews's articles from the Sierra Maestra made Castro into a worldwide legend, and the legend only grew as Castro continued to survive and pile up victories. When, at last, Batista fled Havana on New Year's Day, many Americans were as thrilled as the exultant Cuban masses. An evil man had been deposed; a new man, young and idealistic and charismatic, had won the hearts and minds of much of the world.

And now here he was in the flesh, stepping onto the tarmac into a crowd of U.S. officials.

"Viva Castro!"

Before Castro could shake hands with Roy Rubottom, the State Department official there to greet him, a hundred or so fans suddenly rushed through a ring of police. Castro received the swarm of adulation with handshakes and hugs, then wended his way to a thicket of microphones. Generally, microphones inspired him to long and rambling orations. Not tonight. His voice was hoarse and soft, and surprisingly high-pitched for so large a man.

"I have come here to speak to the people of the United States," he began in halting English. "I hope the people of the United States will understand better the people of Cuba, and I hope to understand better the people of the United States."

He turned for the limousine parked on the tarmac, then strode right past it to a large crowd shouting from the other side of a chain-link fence. His security detail, including local police officers and forty agents from the State Department's Division of Physical Security, scrambled to keep up. The State Department had already fielded numerous threats against Castro and had every reason to worry an assassin might try to gun him down. But the greatest danger to Castro, evidently, was going to be Castro himself. "He must be crazy," one of the guards observed as Castro flung himself at the crowd.

CRAZY WAS A common assessment of Fidel Castro in certain quarters of the American government. Few Americans were sorry to see Batista go, and the United States had quickly recognized the new regime. But some officials, such as Roy Rubottom, the assistant secretary of state Castro left standing on the tarmac, were already expressing grave doubts about the new prime minister. Both CIA intelligence and firsthand reports from Cuba suggested that Castro was erratic, tyrannical, and bloodthirsty. Since arriving in Havana at the start of the year to take the reins of government, Castro had either ordered or allowed the executions of more than five hundred Batista supporters. Between executions, he'd delivered stupendous diatribes, some lasting as long as three or four hours and many of them laced with anti-American sentiments. American conservatives such as Senator Barry Goldwater were particularly alarmed by the tone of the new Cuban leader, but even the liberal and outspokenly pro-Castro New York congressman Adam Clayton Powell had returned from a March visit to Cuba with alarming tales of a man who slept a couple of hours a night, kept himself awake with high doses of Benzedrine, fell frequently into incoherence when talking, and had, in general, "gone haywire."

The Fidel Castro who arrived in the United States in April 1959 may have been sleep-deprived and criminally indifferent to his own safety, but he was not noticeably incoherent or haywire. On the contrary, he struck most of those who met him during his eleven-day visit to America as reasonable and amiable, even charming.

Among those who would not get a chance to experience Castro's charms firsthand was Dwight Eisenhower. The president had excused himself from meeting the Cuban revolutionary by decamping to Augusta, Georgia, to play golf. Eisenhower was not kindly disposed to revolutionaries in the first place; moreover, he was irritated by the circumstances that brought Castro to Washington. Generally, a foreign head of state would not think to visit America without an official invitation from the State Department. Castro had come, instead, by invitation of the American Society of Newspaper Editors (ASNE) to deliver the keynote address at the organization's annual meeting on April 17. In Castro's defense (and ASNE's), he had accepted the invitation before he officially became Cuba's prime minister. Still, it was an unseemly breach of protocol for him to show up like this, and Eisenhower was not pleased.

As it happened, Castro's arrival in the United States came on a very difficult day for Eisenhower. That morning, the president had learned in a phone call that his longtime secretary of state, John Foster Dulles, was resigning, effective immediately. For eight years, Dulles had been the ballast, if not the rudder, of Eisenhower's anti-Communist foreign policy. Now he was in Walter Reed Army Hospital with terminal abdominal cancer. That John Foster Dulles should end his career on the very day Fidel Castro landed in America is one of those coincidences that would seem, like so many others of the next two years, to have been plotted by a roomful of cackling Soviet scriptwriters bunked up in a commune near the Kremlin.

THE MORNING AFTER his arrival, Castro awoke in a bedroom in the Cuban embassy on Sixteenth Street and began to practice his English.

An aide ran out to buy him a comb and toothbrush. Castro was usually indifferent to personal hygiene, but he was eager to make a good impression on the Americans. He even managed to arrive on time for a luncheon later in the day with Acting Secretary of State Christian Herter. The men chatted cordially, then exchanged toasts. Castro bounced out of the lunch beaming. Herter, cane in hand, emerged sober but impressed. After the lunch, Herter informed President Eisenhower that he found Castro to be "a child in many ways, quite immature regarding problems of government." A few days later, though, a State Department memorandum described the Cuban prime minister as "a man on his best behavior."

This was, as far as it went, an accurate appraisal. Before coming to America, Castro had wired a New York City public relations firm to advise him. In addition to urging Castro to smile a lot, the publicists suggested, less astutely, that he shave his beard to adopt a clean-cut appearance. Castro wisely ignored the latter tip. The beard was part of his mystique. People *loved* the beard. Indeed, popular novelty items in America that spring were fake Castro beards woven from treated dog or fox hair. "When we finish our job," Castro told one interlocutor, "we will cut off our whiskers."

The most sensible advice the public relations executives gave Castro seems to have been this: *tell them what they want to hear.* And what they wanted to hear in the spring of 1959—"they" being government officials, the American press, and the public—was that Fidel Castro was not a Communist.

"What is your connection with communism, if any?" asked Senator Alexander Wiley of Wisconsin when the Cuban visited the U.S. Capitol on Friday morning.

"None," Castro replied, then went on to repeat variations of the answer for the next hour and a half. Afterward, senators and congressmen pronounced themselves cautiously satisfied. "I feel reassured about a number of matters I've been concerned about in Cuba," said Senator Russell Long of Louisiana. Representative James Fulton, a Republican from Pennsylvania, pronounced Castro an "*amigo nuevo.*" Even Senator George Smathers of Florida, a persistent critic of Castro, came away impressed. He remained convinced that Castro's government was "peppered" with Communists, but the prime minister, Smathers acknowledged, appeared to be a "good man."

Everywhere the question was the same: *Are you a Communist, Dr. Castro?* Have you ever been a Communist, or do you sympathize with Communists? Everywhere Castro gave the same answer: *No, he was not a Communist.* Never had been. Never would be.

CASTRO'S TRUE IDEOLOGY in April 1959 is, even now, difficult to pin down. Certainly there were Communists around him and close to him, men who had fought alongside him in the Sierra Maestra and now served in his government and army. His brother Raúl and his chief adviser, Ernesto "Che" Guevara, both had strong ties to the Communist Party. Most informed American observers, though, concluded that Castro was telling the truth when he said he was not a Communist. He had never been in the Communist Party, and while he welcomed Communists into his revolution, he welcomed Cubans of other political stripes, too. As for his relationships with Communists in other countries, he had none. His first known contact with any Soviet official, in fact, came during his visit to Washington that April, when he exchanged pleasantries with the Soviet ambassador during a reception at the Cuban embassy.

A more interesting question than *when* Fidel Castro became a Communist is *why* Americans were so intent on divining his ideological affiliation. Twenty-first-century Americans may find the obsession with Castro and communism slightly bizarre, if not hysterical. Americans who lived through the darkest days of the Cold War, though, will recall the issue as absolutely essential, even downright *existential.* The spread of communism was the defining geopolitical concern of the age—the organizing principle on which nearly every act and policy of U.S. foreign relations depended.

There was, in fact, plenty to fear. State-sponsored communism had risen with startling swiftness after the close of the Second World War. Russia and China were in Communist hands and apparently conspiring to spread their creed to every hill and dale on the globe. And there was no shortage of places ripe for cultivation. As European powers pulled out of former colonies, struggling new nations in Africa, Asia, and the Middle East saw in communism an ideology that offered a way out of imperialist shackles. President Eisenhower famously described the spread of communism from nation to nation as a domino effect. A more apt analogy might be germ theory. As each new country succumbed to communism, it would presumably infect its neighbor, which in turn would infect *its* neighbor, and so on, until the contagion was forcibly checked—or until Communists ruled the world. Communists were not modest about their viral creed. Sooner or later, Khrushchev had infamously warned the West, "We shall bury you."

Schoolyard taunts were typical of the Soviet premier, but Americans could not easily dismiss them. They had lived through the Second World War and watched Adolph Hitler actively seek world domination. They had every reason to believe that Khrushchev and Mao Tse-tung were seeking it, too—and plenty of evidence the Communists had the tools to succeed. The same year China went Red, 1949, the Soviet Union had exploded an atomic bomb, wiping away in a single detonation America's monopoly on the most destructive force ever unleashed by humans. Eight years after that, in 1957, the Soviets had leaped ahead in space technology with Sputnik, the first satellite ever launched into orbit. Sputnik was a stunning blow to American morale, but more important than the satellite itself was the rocket the Soviets had used to launch it. If they could make a rocket with enough thrust to carry a 184-pound aluminum sphere to the edge of Earth's atmosphere, it stood to reason that they could produce rockets capable of delivering nuclear-tipped intercontinental ballistic missiles to the heart of the United States.

Not only did the Soviets lead in missile thrust, they were ahead in missile numbers, too. Or so it was feared. According to conventional wisdom at the time, they owned a significant advantage in intercontinental missiles, as high as three-to-one. The "missile gap" would later be exposed as a canard, but not before it induced many worrisome headlines and sleepless nights. *Newsweek* put it this way at the start of 1959: "If the intercontinental ballistic missile and the thermonuclear warhead are the ultimate weapons which man has devised for destruction, then the forces of Premier Nikita Khrushchev are unquestionably ahead." In a blaring rhetorical headline *U.S. News & World Report* asked, HOW READY IS U.S. FOR WAR? The answer, the magazine informed its readers, was: *Not very.*

Beyond very real and definable anxieties regarding nuclear obliteration, a vaguer insecurity seemed to grip many Americans of the 1950s. This was reflected in edgy bestsellers such as *The Man in the Gray Flannel Suit* and *The Ugly American*, in which American might, for all its obvious fecundity and influence, was portrayed as a tenuous and at some level hollow proposition. Richard Nixon's so-called kitchen debate with Nikita Khrushchev in the summer of 1959, in which the two men squabbled about the respective merits of their nation's kitchen appliances, would be hailed as a victory for Nixon's debating skills and a rousing defense of American prosperity and industry, but in the end Nixon's argument was strikingly vapid. This was America's great claim? That it had better dishwashers?

Of course, the larger question was: why, if Americans were so sure of the superiority of their economic system over communism, were they not more confident in its ultimate triumph? "Yours is a great country," Castro would tell Nixon when they met during his visit to Washington that April. "Your people, therefore, should be proud and confident and happy. But everyplace I go you seem afraid." Another bestseller in America that spring supported his observation. Titled *What We Must Know about Communism*, the book was a scarifying survey of the Communists' growing reach and ambition. "We have written this book because we had to," the authors Harry and Bonaro Overstreet declared in the book's introduction. "There comes a point when the world's peril turns into every individual's responsibility."

What all of the fear meant, reasonable or not, was that the specter of a Communist country ninety miles from America was intolerable. The beautiful island nation that America had liberated from Spain and supported economically for decades simply could not be a vassal and beachhead of the international Communist conspiracy. And so the question had to be asked: *Dr. Castro, are you a Communist?*—and Dr. Castro gave the right answer: *No, I am not a Communist.*

AS IF TO drive home Americans' fears of nuclear-armed Communists, the afternoon of Castro's second full day in Washington, April 17, happened to coincide with a peculiar national rite of mid-twentieth-century America known as Operation Alert. The sixth annual drill organized by the Office of Civil and Defense Mobilization, Operation Alert, 1959, was meant to prepare Americans for nuclear attack. The drill postulated a bombardment by a fleet of enemy aircraft (presumably of Soviet origin) that had been spotted flying over northern Canada, heading for the United States. For a few hours that spring day, Americans were supposed to imagine they were under nuclear attack.

The first sign of the mock attack came at 11:30 a.m. in the East—10:30 in Chicago, 8:30 on the West Coast—when television and radio programming was replaced by urgent warnings broadcast on dedicated frequencies. At 12:30 p.m. Central Time, air raid sirens sounded a second warning in Chicago. Moments later, an imaginary ten-megaton hydrogen bomb landed at Sixty-third Street and Kedzie Avenue. According to the drill's specifications, 229,625 Chicagoans died instantly in the blast zone and another 622,284 were severely injured. Hundreds more make-believe bombs rained over the nation that afternoon. In New York City, where participation in Operation Alert was compulsory, lunchtime traffic came to a halt. Times Square cleared out and became a "sunlit wasteland," according to the *New York Times*, as New Yorkers sought shelter in subway stations and stores, restaurants and bars. "People ate and drank and looked at the empty streets from behind plate glass," the *Times* reported. IT WAS SO QUIET, wisecracked a *New York Daily News* headline, YOU COULD HEAR A BOMB DROP.

Washington, D.C., too, was targeted in Operation Alert, but no one there seemed to pay the drill any mind. As the capital went up in imaginary flames outside, Fidel Castro stood at a podium inside the ballroom of the Statler-Hilton Hotel, delivering his keynote address to the American Society of Newspaper Editors. "I have said very clearly that we are not Communists," he told the audience in full-throated English. "Our revolution is a humanistic one." Castro gave the editors a litany of reassurances bound to ring pleasantly in American ears. Though he would legally expropriate some privately owned lands, he did not intend to confiscate American property as part of his agrarian reform program, he informed them. Free elections were on the way. As for a free press, not to worry—he cherished it as "the first enemy of dictatorship." The editors had greeted Castro with tepid claps, but now, as he finished, they applauded enthusiastically.

CASTRO WENT ON to enjoy a few more hectic days in Washington. Grinning cheerfully and grappling gamely, if not always successfully, with the English language, he entertained rowdy guests at the Cuban embassy and dashed around town in a siren-blaring motorcade. Everywhere he went he attracted excitement and admiring glances. "He has such kind eyes," one woman observed of him. "Doesn't he remind you of a younger Jimmy Stewart?" asked another. He stopped at a school playground to play peekaboo with small children. He signed autographs and lurched irrepressibly toward any friendly mob and its potentially homicidal embrace. On Saturday evening he gave his already agitated security detail the slip and went out for a midnight snack of Chinese food. "Go get me a tent," an exasperated agent of the State Department's security force was overheard muttering under his breath. "I got everything else for this circus."

On a drizzly April 19, following visits to Mount Vernon and the Lincoln and Jefferson memorials, Castro returned to Capitol Hill to meet Richard Nixon. The vice president escorted Castro through the Capitol, quiet on this Sunday afternoon, to his office. For two and a half hours, Castro and Nixon spoke. The meeting was private, but six days later Nixon sent a classified memorandum about it to the president and several other members of the administration. This memorandum remains the best record of the meeting and a curious, possibly seminal, document in the history of the Bay of Pigs.

According to Nixon's memo, Castro arrived at the Capitol "somewhat nervous and tense," concerned he had performed poorly on *Meet the Press* earlier in the day. After reassuring his guest he'd done just fine, Nixon lost no time in lecturing Castro about the value of free elections, habeas corpus, and other fine points of democracy. "I frankly doubt that I made too much impression upon him but he did listen and appeared to be somewhat receptive," wrote Nixon. The conversation turned, inevitably, to communism. Nixon had a long record as a Communist-buster, going back to his attacks on Alger Hiss in the 1940s, and was just the man to shake the red off a young rebel. As Nixon harangued Castro about the dangers of communism, Castro grew irritated—"This man has spent the whole time scolding me," he later told an aide—but remained polite, if not quite solicitous.

The Nixon-Castro meeting is intriguing not only for what transpired but also for Nixon's depiction of it later. In his 1962 book *Six Crises,* Nixon would describe his encounter with Castro as the turning point in his view of the Cuban leader. Quoting the memo he sent to Eisenhower, Nixon wrote that he "stated flatly that I was convinced Castro was 'either incredibly naïve about Communism or under Communist discipline' and that we would have to treat him and deal with him accordingly—under no further illusions. . . ." Nixon claimed that he at once became the administration's chief advocate for overturning Castro.

Actually, Nixon's original memorandum, which became public only twenty years after he wrote it, belies his own description of the document. Although Nixon did state in his memo that Castro was either "naïve" or "under Communist discipline," he also added that "my guess is the former" and that, overall, his impression was "somewhat mixed." Not exactly warm praise, but hardly a decisive call to arms. Richard Nixon apparently shared the ambivalence held by most of official Washington toward Fidel Castro in April 1959. The only thing he knew for sure was that Castro possessed "those indefinable qualities which make him a leader of men" and that "we have no choice but at least to try to orient him in the right direction."

New York, April 1959

ON MONDAY, APRIL 20, Castro boarded a private rail car at Washington's Union Station and entrained up the eastern seaboard to Princeton, New Jersey, where he stopped to address a seminar of mostly adoring Princeton University students. The following morning, Castro's caravan sped down Route 206 to the Lawrenceville School, one of the nation's oldest and most prestigious boarding academies. Wearing a long, dark trench coat over his fatigues and clenching a cigar between his teeth, Castro entered the school's stone chapel to address six hundred boys in jackets and ties. The political philosophy of Fidel Castro was probably not on the curriculum most of the boys' parents envisioned when they sent their sons to Lawrenceville; nor was this ivy-clad chapel a natural habitat for the rebel of the Sierra Maestra. But the boys greeted Castro with "thunderous applause," according to the school newspaper, and Castro, for his part, said nothing to offend. "I feel something sad of not knowing well the English to express my emotion," he told the boys apologetically. "I cannot speak long here for two reasons. One, because the train is waiting and I have a large program and here in the United States somebody have taught me to be punctual. Second, because my English this morning didn't woke up very clear."

It was in New York City that Fidel Castro took his circus to the zoo. The chaos began the moment he stepped out onto the concourse level of Pennsylvania Station on the afternoon of April 21. Twenty thousand screaming well-wishers packed the train station and spilled for blocks onto Seventh Avenue. A scrum of police tried to usher Castro quickly through the crowd, but he was not to be denied. "I must see the people!" he called out, breaking through the security perimeter to clutch hands and return embraces. It took half an hour to maneuver him through the crowd to his hotel, the Statler, directly across Seventh Avenue.

The next four days passed in a whirl of press conferences and lectures, of meetings and interviews, Castro beaming through most of it like a man on his second honeymoon. He visited Columbia University, toured City Hall, rode to the top of the Empire State Building, and shook hands with Jackie Robinson. Wherever he went he was besieged by photographers and reporters. *Life* even caught him in his hotel one morning, tousled from sleep and wearing striped pajamas.

There were private moments, too. One of these occurred behind closed doors at the Statler, where Castro was interviewed by a CIA agent. The interview had been arranged by Castro's minister of finance, Rufo López-Fresquet, a politically moderate economist who hoped to prove that his boss was no Communist. "We shall bestow on him the fictitious name of 'Mr. Frank Bender,'" López-Fresquet later wrote of the CIA agent who came to visit Castro. The name was a giveaway. Frank Bender was the alias of a German-born CIA veteran named Gerry Droller. One American official had described Droller to López-Fresquet as "the highest authority of American intelligence on the Communists in Latin America." This rather overstated the case. Droller did work on the Latin America desk of the CIA, but an "authority" he was not. For one thing, he spoke no Spanish. His own colleagues at the CIA tended to dismiss him as a know-it-all who blew a lot of smoke—literally. He had a passion for cigars that rivaled Castro's.

Maybe it was the cigars that got to Droller's head. In any case, he came out of the three-hour meeting in a state of near intoxication. "Castro is not only not a Communist," he exclaimed to López-Fresquet, "but he is a strong anti-Communist fighter."

A year later, Gerry Droller, alias Frank Bender, would be working with a task force at the CIA to remove Fidel Castro from power and, if possible, eliminate him from the face of the Earth.

AMONG THE IRONIES attending Castro's 1959 trip to America were the great lengths to which U.S. federal security agents and local police went to keep him alive. No visitor to America had ever received such lavish protection. Few had ever disdained it so cavalierly. What had been true in Washington was doubly so in New York. From the moment Castro arrived late Tuesday morning, he was surrounded by concentric rings of federal agents, plainclothesmen, and uniformed police officers, all of whom he treated with a mix of bemusement and benign neglect. "The hell of it is you never know when, where or how he's going," a police officer told the *New York Post*. "He just decides every once in a while to go for a walk and talk to people."

The police stepped up their already extraordinary measures as the week progressed and threats on Castro's life proliferated. The most picturesque of the reported plots against him had five brothers traveling from Chicago in a black and white 1957 Chevrolet with Florida license plates. 5 HUNTED IN CASTRO DEATH PLOT, the *Post* blared on its front page on Thursday, April 23. The brothers were presumably seeking vengeance on behalf of organized crime. Plenty of people wanted Castro dead for political reasons, but for the mob it was strictly business: one of Castro's first acts as Cuba's leader had been to shut down the mob-run Havana casinos.

The five brothers were soon tracked to Philadelphia, where it turned out they were engaged not in a Castro death plot but in honest labor. No sooner was this threat resolved, however, than a new one surfaced: now two men were speeding east from Detroit in a "dirty gray" Cadillac with Michigan plates. As Port Authority police kept a close watch on incoming lanes of bridges and tunnels, enterprising journalists tracked down Meyer Lansky in Florida. The infamous mobster and former dean of the Havana mob was living in financial ruin, thanks largely to Castro. Lansky refused to speak to reporters, but his wife took the phone. "It's a lie," she said of a mob plot. "It's so ridiculous there is no answer." Another mobster with Havana ties, Joe "Doc" Stracher, was reached in Las Vegas. "What plot? I've got nothing to discuss," said Stracher. "Forget about it."

The drama came to a head on Friday, April 24. Castro was scheduled to give a speech in Central Park that evening, by far his largest venue yet. Police urged him to call it off in the interest of self-preservation and public safety, but Castro refused. Before the speech, Castro relaxed with an impromptu visit to the Bronx Zoo. As he and his entourage sauntered around the zoo, passing astonished mothers and gaping children, Castro seemed to enjoy himself immensely. He fed potatoes and carrots to the elephants. He offered sugar to a gorilla. He ate a hot dog and rode on a miniature electric railroad. And then, to the dismay of everyone present—especially those charged with keeping him alive—he leaped over a protective railing and reached his fingers through the cage to pat a Bengal tiger on the cheek. "This is like prison," said Castro, sympathizing with the tiger. "I have been to prison, too."

* * *

CASTRO WAS STILL at the zoo when crowds began to gather in C
Park. By four-thirty, the area in front of the band shell was throng
six, police began to muster. The NYPD brought in nearly a thou
officers, including dozens on horseback. Lookouts were posted on C
tral Park West rooftops to watch for snipers. Powerful searchlig
scanned the trees and "flickered over the scene like heat lightning
according to the *New York Times*, "turning the leaves pale violet and bri
liant green and the trunks of the trees a luminous white."

Castro arrived at eight-thirty behind a motorcycle escort, stepping
onto the band shell stage to cheers and shouts of *"Viva Castro!"* He
addressed the crowd in Spanish for two hours under a fair night sky.
Occasionally, a roving searchlight flashed over him and he shaded his
eyes with a hand. He spoke of Cuba's aspirations and praised the United
States for its understanding. Whatever hard feelings he'd had for his
neighbor to the north seemed to have softened. As he had put it in
English in a speech earlier that day to the National Press Club, he would
return to Cuba with a "stronger faith" in the bond of friendship.

He was still speaking when a scuffle broke out behind the band shell.
Two policemen had come upon a young man lurking on the slope back
there. When they searched his belongings they found a bomb manufactured from a footlong section of a vacuum cleaner handle, filled with a
mixture of sulfur and zinc. The young man told the police that he had
come to the park "looking for excitement." Castro, unaware of the commotion behind him, went on talking for another fifteen minutes.

He left the city the next morning, just as he had arrived, by train and
surrounded by thousands of people. "Thank God that's over," exhaled a
cop as the train pulled away. The police were happy to see him go, but
few could deny the visit had been a success. "He made it quite clear that
neither he nor anyone of importance in his Government so far as he
knew was Communist," concluded an editorial in the *New York Times*.
"By the same token it seems obvious that Americans feel better about
Castro than they did before."

Within months, the U.S. government, having just spent millions of
dollars and employed thousands of men to protect Fidel Castro from
harm, would be taking the first steps to remove him from power by
whatever means necessary.

ABOUT THE AUTHOR

...asenberger is the author of *The Brilliant Disaster: JFK, Castro, and ...erica's Doomed Invasion of Cuba's Bay of Pigs* and *High Steel: The Dar...g Men Who Built the World's Greatest Skyline.* He has been a contributing editor to *Vanity Fair* and writes frequently for *The New York Times.* He lives in Manhattan with his wife and sons.

* * *

CASTRO WAS STILL at the zoo when crowds began to gather in Central Park. By four-thirty, the area in front of the band shell was thronged. At six, police began to muster. The NYPD brought in nearly a thousand officers, including dozens on horseback. Lookouts were posted on Central Park West rooftops to watch for snipers. Powerful searchlights scanned the trees and "flickered over the scene like heat lightning," according to the *New York Times*, "turning the leaves pale violet and brilliant green and the trunks of the trees a luminous white."

Castro arrived at eight-thirty behind a motorcycle escort, stepping onto the band shell stage to cheers and shouts of *"Viva Castro!"* He addressed the crowd in Spanish for two hours under a fair night sky. Occasionally, a roving searchlight flashed over him and he shaded his eyes with a hand. He spoke of Cuba's aspirations and praised the United States for its understanding. Whatever hard feelings he'd had for his neighbor to the north seemed to have softened. As he had put it in English in a speech earlier that day to the National Press Club, he would return to Cuba with a "stronger faith" in the bond of friendship.

He was still speaking when a scuffle broke out behind the band shell. Two policemen had come upon a young man lurking on the slope back there. When they searched his belongings they found a bomb manufactured from a footlong section of a vacuum cleaner handle, filled with a mixture of sulfur and zinc. The young man told the police that he had come to the park "looking for excitement." Castro, unaware of the commotion behind him, went on talking for another fifteen minutes.

He left the city the next morning, just as he had arrived, by train and surrounded by thousands of people. "Thank God that's over," exhaled a cop as the train pulled away. The police were happy to see him go, but few could deny the visit had been a success. "He made it quite clear that neither he nor anyone of importance in his Government so far as he knew was Communist," concluded an editorial in the *New York Times*. "By the same token it seems obvious that Americans feel better about Castro than they did before."

Within months, the U.S. government, having just spent millions of dollars and employed thousands of men to protect Fidel Castro from harm, would be taking the first steps to remove him from power by whatever means necessary.

ABOUT THE AUTHOR

Jim Rasenberger is the author of *The Brilliant Disaster: JFK, Castro, and America's Doomed Invasion of Cuba's Bay of Pigs* and *High Steel: The Daring Men Who Built the World's Greatest Skyline*. He has been a contributing editor to *Vanity Fair* and writes frequently for *The New York Times*. He lives in Manhattan with his wife and sons.